BASIC INTRODUCTION TO
BIOELECTROMAGNETICS

SECOND EDITION

BASIC INTRODUCTION TO
BIOELECTROMAGNETICS

SECOND EDITION

CYNTHIA FURSE
DOUGLAS A. CHRISTENSEN
CARL H. DURNEY

CRC Press
Taylor & Francis Group
Boca Raton   London   New York

CRC Press is an imprint of the
Taylor & Francis Group, an **informa** business

CRC Press
Taylor & Francis Group
6000 Broken Sound Parkway NW, Suite 300
Boca Raton, FL 33487-2742

© 2009 by Taylor & Francis Group, LLC
CRC Press is an imprint of Taylor & Francis Group, an Informa business

International Standard Book Number-13: 978-1-4200-5542-9 (Hardcover)

Library of Congress Cataloging-in-Publication Data

Furse, Cynthia, 1963-
   Basic introduction to bioelectromagnetics. -- 2nd ed. / Cynthia Furse.
      p. ; cm.
   Rev. ed. of: Basic introduction to bioelectromagnetics / Carl H. Durney, Douglas A. Christensen. c2000.
   Includes bibliographical references and index.
   ISBN 978-1-4200-5542-9 (hardcover : alk. paper)
   1. Electromagnetism--Physiological effect. 2. Electromagnetic fields. I. Durney, Carl H., 1931- Basic introduction to bioelectromagnetics. II. Title.
   [DNLM: 1. Electromagnetics. 2. Biophysics. QT 34 F991b 2008]

QP82.2.E43D87 2008
612'.01442--dc22   2008044157

Visit the Taylor & Francis Web site at
http://www.taylorandfrancis.com

and the CRC Press Web site at
http://www.crcpress.com

For Katie

For Laraine

For Marie

Contents

Preface

While doing research in bioelectromagnetics (the interaction of electromagnetic fields with biological systems) for more than 30 years, we have sensed the need some life scientists have to understand the basic concepts and characteristic behaviors of electromagnetic (EM) fields so they can work effectively with physicists and electrical engineers in interdisciplinary research. Because most EM books are based heavily on vector calculus and partial differential equations, however, little written information about EM fields is available to satisfy this need. Many times over the years life scientists have asked us for references to EM books appropriate for them, but we could give none. These scientists wanted to understand how the fields worked and what controlled them, what factors were important in experimental setups and which were not. They had a great deal of curiosity in how fields were applied in their specific medical application. Yet they rarely, if ever, actually computed the fields themselves. These scientists needed a strong intuitive understanding of electromagnetic fields. We wrote the first edition of this book in an effort to fulfill that need, as well as to help others who want to learn about electromagnetics, but do not have the mathematical background to understand typical books on electromagnetics. The second edition of this book continues in that vein. The material is rearranged in many places to give the reader the details "just in time" to understand the applications. The second edition is also augmented by over forty medical applications of electromagnetics commonly found in clinical settings as well as a new and expanded Chapter 6 describing emerging methods and understanding about bioelectromagnetics. These applications are used to illustrate the basic principles in this book and how they are translated into real devices.

The purpose of this book is to explain the basic concepts, fundamental principles, and characteristic behaviors of electric and magnetic fields to those who do not have a background in vector calculus and partial differential equations. In particular, it is intended for life scientists collaborating with engineers or physicists in work involving the interaction of electromagnetic fields with biological systems. It should also be helpful to health physicists, industrial hygienists, and public health workers concerned with possible hazards or beneficial applications of electromagnetic field exposure, and those concerned with magnetic resonance imaging, implantable medical devices, electrophysiology, optical interactions with tissue, wireless communication devices, and more. Futhermore, this book may also be useful to traditional electrical engineers and physicists who are learning or have already learned the calculus-based mathematical calculations associated with traditional electromagnetics but who would like to have a stronger intuitive grasp of the subject.

In stark contrast to typical EM books that require a background in vector calculus and partial differential equations, this book requires only a background in algebra (some acquaintance with trigonometric functions would also be helpful), but it explains in detail the basic concepts, fundamental principles, and characteristic behaviors of EM fields using pictures, field maps, and graphs and numerous real-world applications. The explanations include a minimum of mathematical relationships, with the emphasis on qualitative behaviors and graphical descriptions. Nevertheless, in spite of the de-emphasis on advanced mathematics, the concepts of EM field theory are still treated in a comprehensive and accurate manner. The material covers the entire frequency spectrum from direct current (DC) up through optical frequencies. Practical explanations are given to help readers understand real situations involving EM fields. Over two hundred illustrations are included to augment qualitative explanations.

The first chapter gives an introduction to the fundamentals of EM field theory, and explains how characteristic behaviors can be effectively grouped in three categories defined by the wavelength of the EM fields compared to the size of the objects with which they interact: (1) when the wavelength is much larger than the size of the objects, (2) when it is about the same, and (3) when the wavelength is much smaller than the size of the objects. Chapters 2 to 4 respectively explain the characteristic behaviors in each of these three categories and how they are applied to applications in those frequency bands. Chapter 5 explains some of the principles of EM fields that are quantified in detailed and complex environments typical of bioectromagnetic applications. This calculation of the doses of the electromagnetic fields is called *dosimetry*. The book concludes with Chapter 6, which discusses the emerging and future applications of bioelectromagnetics.

We sincerely hope that this book will be useful (and enjoyable!) for its intended readers. We welcome comments and suggestions for improving it.

Cynthia M. Furse
Professor of Electrical Engineering
University of Utah

Douglas A. Christensen
Professor of Electrical Engineering
Professor of Bioengineering
University of Utah

Carl H. Durney
Professor Emeritus of Electrical Engineering
Professor Emeritus of Bioengineering
University of Utah

Authors

Cynthia M. Furse was born in Stillwater, Maine on May 7, 1963. She received her BSEE degree in 1986, her MSEE in 1988, and her PhD in electrical engineering in 1994, all from the University of Utah. She was an NSF CISE Fellow at the University of Utah from 1994–1997, where she developed computational methods for determining the absorption of electromagnetic fields in the head from cellular telephones. She then was an assistant and associate professor of electrical engineering at Utah State University, where she taught electromagnetics, wireless communication, antennas, numerical electromagnetics, microwave engineering, and project management. While at USU, she established the Center of Excellence for Smart Sensors to create and commercialize sensors for evaluating complex environments such as the human body, underground geophysical phenomenon, and wiring systems in aircraft. She was also the director of the Richard and Moonyeen Anderson Wireless Teaching and Research Center. Dr. Furse was the Professor of the Year in the College of Engineering at Utah State University for the year 2000 and the Faculty Employee of the Year 2002. In 2002 she moved to the University of Utah, where she is now a professor of electrical engineering. Dr. Furse's major biological research interests include telemetry systems for the human body, simulation of fields in the body, and coil designs for biological imaging. She is also the chief scientist for LiveWire Test Labs, Inc., a university spinoff company commercializing sensors for locating intermittent faults on live aircraft wiring. She received the Distinguished Educator award in the College of Engineering and also the Distinguished Young Alumni award in the Department of Electrical and Computer Engineering in 2008. She is active in K-12 outreach programs to expose young people to the excitement of engineering.

Dr. Furse is a member of the IEEE where she was elected Fellow in 2008, Commission K of the Union of Radio Science International (URSI), Phi Kappa Phi, Eta Kappa Nu, Tau Beta Bi, the American Society of Engineering Education, the Society of Women Engineers, and the Applied Computational Electromagnetics Society. She is the past-chairman (1999–2007) of the IEEE Antennas and Propagation Society Education Committee and member of the IEEE AP Administrative Committee, editor-in-chief of the *International Journal of Antennas and Propagation*, and founding member of the editorial board of the *Journal of Smart Structures and Systems*. She has also served as a member of the editorial board of the *IEEE Transactions on Antennas and Propagation*, the *Journal of the Applied Computational Electromagnetics Society*, and the *IEEE Applied Wireless Propagation Letters*.

Carl H. Durney was born in Blackfoot, Idaho, on April 22, 1931. He received a BS degree in electrical engineering from Utah State University in 1958, and MS and PhD degrees in electrical engineering from the University of Utah in 1961 and 1964, respectively.

From 1958 to 1959 he was an associate research engineer with Boeing Airplane Company, Seattle, Washington, where he investigated the use of delay lines in control systems. He has been with the University of Utah since 1963, where he is presently professor emeritus of electrical engineering and professor emeritus of bioengineering. From 1965 to 1966, he worked in the area of microwave avalanche diode oscillators at Bell Telephone Laboratories, Holmdel, New Jersey, while on leave from the University of Utah. He was visiting professor at the Massachusetts Institute of Technology doing research in nuclear magnetic resonance (NMR) imaging and hyperthermia for cancer therapy during the 1983–84 academic year while on sabbatical leave from the University of Utah. At the University of Utah, until he retired in 1997, he taught and did research in electromagnetics, engineering pedagogy, electromagnetic biological effects, and medical applications of electromagnetics.

Dr. Durney is or has been a member of IEEE, the Bioelectromagnetics Society, Commissions B and K of the International Union of Radio Science (URSI), Sigma Tau, Phi Kappa Phi, Sigma Pi Sigma, Eta Kappa Nu, and the American Society for Engineering Education (ASEE). He served as vice president (1980–81) and president (1981–82) of the Bioelectromagnetics Society, as a member (1979–88) and chairman (1983–84) of the IEEE Committee on Man and Radiation (COMAR), as a member of the American National Standards Institute C95 Subcommittee IV on Radiation Levels and/or Tolerances with Respect to Personnel (1973–88), as a member of the editorial board of *IEEE Transactions on Microwave Theory and Techniques* (1977–97), and as a member of the editorial board of *Magnetic Resonance Imaging* (1983–95). He was a member of the National Council on Radiation Protection and Measurements from 1990 to 1996. He served as a member of the Peer Review Board on Cellular Telephones (Harvard Center for Risk Analysis) from 1994 to 1997. In 1980, Dr. Durney received the Distinguished Research Award, and in 1993 the Distinguished Teaching Award from the University of Utah. In 1982, he received the ASEE Western Electric Fund Award for excellence in teaching, and the Utah Section IEEE Technical Achievement Award. Utah State University named him College of Engineering Distinguished Alumnus in 1983. In 1990 the Utah Engineering Council named him Utah Engineering Educator of the Year. He was elected a fellow of the IEEE in 1992. In 1993 the Bioelectromagnetics Society awarded him the d'Arsonval Medal.

Douglas A. Christensen was born in Bakersfield, California, on December 14, 1939. He attended Brigham Young University in Provo, Utah, graduating with a BS degree in electrical engineering in 1962. He was valedictorian of the College of Engineering. He attended Stanford University in Palo Alto, California, graduating with an MS degree in electrical engineering in 1963. He then pursued a PhD degree in electrical engineering at the University of Utah, Salt Lake City, graduating in 1967. He was awarded a special postdoctoral fellowship from the National Institutes of Health for studying bioengineering, which he took at the University of Washington, Seattle, from 1972 to 1974. In addition, he has pursued research at the University of California at Santa Barbara and at Cornell University, Ithaca, New York.

Dr. Christensen was appointed an assistant professor of electrical engineering at the University of Utah in 1971. He also received an appointment as an assistant professor of bioengineering at the University of Utah in 1974. He was chairman of the Bioengineering Department from 1985 to 1988. He currently is a professor in both departments.

His industrial experience includes Bell Telephone Laboratories, Murray Hill, New Jersey; International Business Machines Corporation, San Jose, California; Hewlett-Packard Company, Palo Alto, California; and General Motors Research Laboratories, Santa Barbara, California. He has also been a consultant for several companies. His research interests range from electromagnetics to optics to ultrasound. He did early work on a fiberoptic temperature probe used for monitoring temperature during electromagnetic hyperthermia and has worked in numerical techniques for electromagnetic applications, mainly using the finite-difference time-domain method, including its use in optics. He authored a textbook titled *Ultrasonic Bioinstrumentation* and has been co-director of the Center of Excellence for Raman Technology at the University of Utah. He has received the Outstanding Teaching Award and the Outstanding Patent Award from the College of Engineering. His recent interests have been in biomedical optics, especially for sensing and imaging applications.

1

Electric and Magnetic Fields: Basic Concepts

1.1 Introduction

Bioelectromagnetics—the study of how electric and magnetic fields interact with the body—is a tremendously exciting field. Electromagnetic fields are all around us: radio and television signals, cellular telephones, fields from power lines and electrical appliances, radar, and more. They are even within our bodies in the endogenous fields that keep our hearts beating, brains thinking, and muscles moving. Electromagnetic fields can see inside of us to diagnose illness, sometimes before we feel it ourselves, in the form of medical imaging, electrocardiography, electroencephalography, and electrophysiological evaluations. They can heal us through therapeutic interventions for cancer, pain control, bone growth, soft tissue repair, electrophysiological stimulation, and more. And they can injure or kill us through lightning strikes, deep electrical burns, and shock.

Electromagnetic fields are already used in numerous medical devices, and the future (read more in Chapter 6) promises ever more detailed and localized diagnostic and treatment methods. Electromagnetic fields may soon help repair or replace damaged nerve pathways, help the blind to see, the deaf to hear, and the paralyzed to walk again. The promise of bioelectromagnetics seems limited only by our imaginations. However, the promise of bioelectromagnetics is very much limited by the physical nature of the fields themselves and how they can be made to interact with the body. The purpose of this book is to help you understand electromagnetic fields and how they interact with the body, how they are created, how they can be measured and evaluated, and how they can be controlled.

This book begins with the field of classical electromagnetics, which stems from the phenomenon that electric charges exert forces on each other. The concepts of electric and magnetic fields are used to describe the multitude of complex bioeffects that result from this basic phenomenon. Although classical electromagnetic (EM) field theory is typically couched in vector calculus and partial differential equations, many of the basic concepts and characteristic behaviors can be understood without a strong mathematical background. The purpose of this book is to describe and explain these basic concepts and characteristic behaviors with a minimum of mathematics, and to show how they are used in a wide variety of bioelectromagnetic applications. In this chapter we explain the basic concepts of electric and magnetic fields as a basis for what follows in the remainder of the book.

1.2 Electric Field Concepts

A fundamental law, Coulomb's law, states that electric charges exert forces on each other in a direction along the line between the charges. Charges with the same sign repel, and charges with opposite signs attract. The magnitude of the force exerted on one charge by

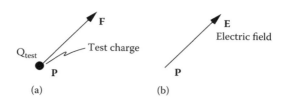

FIGURE 1.1
(a) Force **F** exerted on a charge Q_{test} placed at a point P in space. (b) Electric field **E** at the point P defined as **E** = **F**/Q_{test}.

another charge is inversely proportional to the square of the distance between the two charges. Because keeping track of the forces exerted on individual charges in a complex system of charges is almost impossible in practice, the concept of electric field is used to account for the forces.

The concept of electric field is illustrated by this thought experiment: Place a small test charge Q_{test} at a point in space P, as shown in Figure 1.1(a). Whatever other charges exist will exert a force on this test charge. Measure that force, denoted by **F**. By definition, the *electric field strength* at point P is given by

$$\mathbf{E} = \mathbf{F}/Q_{test} \ (V/m) \tag{1.1}$$

as shown in Figure 1.1(b). (The direction of **E** is in the direction of the force exerted on a positive test charge. The force on a negative test charge, such as an electron, would be in the opposite direction.) Thus, **E** is a force per unit charge. **E** is also called *electric field intensity*, or often just electric field. The units of **E** are volts per meter (V/m).

Because **F** is a vector, **E** is also a vector. A vector is a quantity having both a direction and a magnitude. In this book, vectors are denoted by boldface symbols. The direction of a vector is represented by an arrow, as in Figure 1.1. The magnitude of a vector is represented by the same symbol as the vector, but without boldface. For example, let us define a vector **v** as a velocity having a direction from south to north and a magnitude of 30 meters per second (m/s). Then the magnitude of **v** is expressed as v = 30 m/s. In a similar fashion, E is the magnitude of the vector **E**.

As a consequence of the definition of electric field, a charge Q placed in an electric field **E** will experience a force given by **F** = Q**E**. The larger the **E**, the larger the force **F** exerted on the charge Q. The fundamental effect of an electric field on an object placed in it is to exert forces on the charges in that object, as explained in Section 1.6.

Electric fields are represented graphically in two ways. Figure 1.2 illustrates the first method, using as an example the electric field produced by a single point charge Q. Remember that **E** fields are produced by charges. The **E** produced by a single point charge is perhaps the simplest example of an **E** field. In this first method of displaying **E** fields, the direction of **E** is shown by arrows, and the magnitude of **E** is indicated by the closeness of the arrows. In areas where the arrows are close together, the magnitude is higher than in areas where the arrows are farther apart. For example, near the charge, the arrows are close together, indicating a large E. Farther away from the charge, the arrows are farther apart, indicating a smaller E.

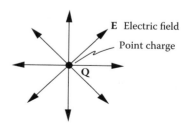

FIGURE 1.2
Plot of the electric field produced by a single point charge Q.

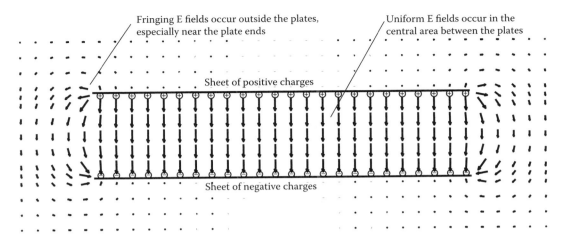

FIGURE 1.3
E field produced by two uniform sheets of charge, positive charge on the top and negative charge on the bottom. This configuration is a representation of a parallel-plate capacitor. The arrows represent the direction of the E field. The length of the arrow indicates the strength (magnitude) of the E field.

The second method of representing vector fields such as E is illustrated in Figure 1.3, which shows the E field produced by two uniform sheets of charge. In this method, the direction of the E field is also shown by arrows. The magnitude of E is indicated by the length of the arrows. The longer the arrow, the larger the E. This second method is often used when the E fields are calculated by numerical methods and plotted by computer graphical methods; this is the method we use most often in this book. The E field produced by the two uniform sheets of charge is uniform near the center of the sheets. At the edges of the sheets, the E bends around, or *fringes*.

Because E fields exert forces on charges, work is required to move a charge from one point in space to another in the presence of an E field. The work done per unit charge is called *electric potential difference*. Electric potential difference is often referred to as potential difference, or just *voltage*, because its unit is the volt (V). When E is known as a function of space, the potential difference between any two points can be calculated. Let us consider

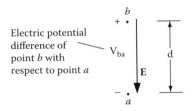

FIGURE 1.4
Configuration for calculating the potential difference of point *b* with respect to point *a* in the presence of E. The charge would move from *b* to *a*.

first the simplest case, when E is uniform in the space between two points, and a positive charge is moved from one point to another along a path in the opposite direction of E, such as moving a charge from point *a* to point *b* in Figure 1.4. For this case, the potential difference of point *b* with respect to point *a* is given by

$$V_b - V_a = V_{ba} = Ed \ (V) \tag{1.2}$$

where d is the distance between the two points. Electrical potential difference refers to potential energy. If a charge were moved from point *a* to point *b*, it would possess potential energy because if it were released, the force produced on it by E would cause it to move, thus converting its potential energy to kinetic energy. When the E field is not uniform, or when the path between *a* and *b* is not exactly in the opposite direction of **E**, Equation 1.2 does not apply, and a more complicated calculation is required. Familiar devices such as 12-volt automobile batteries and 1.5-volt dry cells are used to produce potential differences. Large electric generators produce the potential differences that we use for a multitude of purposes in our homes. Electrocardiograms measure potential differences on the surface of the body caused by the beating heart.

When **E** does not vary with time, or when it varies slowly with time (the frequency is low), the work done in moving charge between two points is independent of the path over which charge is moved between the two points. In this case, the **E** field is said to be *conservative*, and the potential difference is a unique quantity. When **E** varies rapidly with time (the frequency is high), the work done in moving charge between two points generally depends upon the path over which charge is moved between the two points, and a unique potential difference cannot be defined. In this case, **E** is not a conservative field. In special cases (see Section 3.5.1), **E** can vary rapidly with time and still be a conservative field.

Moving charges produce *electric current*, which is defined as the time rate of change of charge. The unit of charge is the coulomb (C). Current at a given point in space is the amount of charge passing that point per second. The unit of current is the ampere (A). Thus, 1 A is equivalent to 1 C/s. *Current density* is defined as current per unit area. Its units are amperes per square meter (A/m^2).

If a time-constant potential difference V is applied between two points and a total current I flows between the two points as a result of this applied voltage, then the current is given by I = V/R, where R is the *resistance* (units are ohms) between the two points. As its name implies, resistance opposes the flow of current. This relationship is called *Ohm's law*. It is one of the fundamental laws of electric circuit theory.

The electric field shown in Figure 1.3 could also be produced by replacing the two sheets of charge with metal plates and applying a potential difference between the two, by connecting, for example, a battery between the plates. The potential difference would produce current through the battery, transferring charge from one plate to the other, thus producing charged plates that would be equivalent to the configuration of Figure 1.3.

HOW ELECTRIC FIELDS ARE MEASURED

Electric fields are measured using metallic antennas. Electric fields (for example, the open lines that travel from positive to negative charges) are picked up by straight antennas, which are oriented parallel to the electric field lines. These straight antennas have a space in the middle that is left open to create a measurable voltage difference. An example is shown in Figure 1.5. This miniature electric field probe antenna was designed for assessment of compliance of electromagnetic devices with radio frequency (RF) exposure guidelines. Measurement of fields in or near the body is difficult, because a metal object (such as a measurement antenna) can perturb the fields. This small dipole antenna was specifically designed to receive the localized fields without perturbing them. This probe picks up electric fields along its axis, but fields oriented in any other direction are ignored. When all three components of the electric field vector are desired (either separately or in combination to find total electric field strength), three perpendicular linear antennas are used, as shown in the probe in Figure 1.6. Each antenna picks up the electric field parallel to its major axis. The three perpendicular electric field vectors can be measured independently or combined to give total electric field.

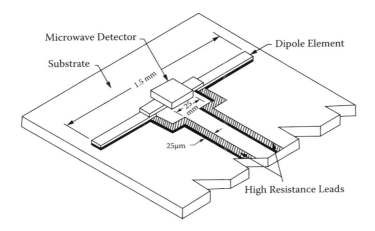

FIGURE 1.5
Miniature printed dipole antenna for measurement of electric fields to determine cell phone RF exposure compliance. (From Bassen, H., and Smith, G., *IEEE Trans. AP*, 31, 710–18, 1983. © 1983 IEEE. With permission.)

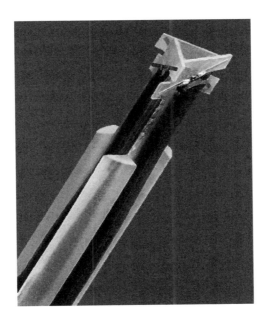

FIGURE 1.6
Electric field probe manufactured by SPEAG. The diameter of the tip is 3.9 mm. (From Schmid & Partner Engineering AG, Zurich. Reprinted with permission.)

1.3 Magnetic Field Concepts

In the previous section, electric field concepts were explained as a means of accounting for the forces between charges that act on a line between the charges. When charges are moving, they exert another kind of force on each other that is not along a line between the

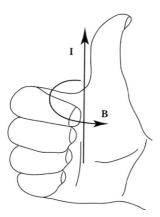

FIGURE 1.7
The right-hand rule can be used to describe the direction of the current and magnetic field. This rule can be used in two ways. First, the thumb can point in the direction of the current (**I**), and the fingers represent the magnetic (**B**) field (as shown). Alternatively, the thumb can point in the direction of the magnetic field, and the fingers will represent the direction of the current that produced it.

charges. Magnetic fields are used to account for this other kind of force. Moving charges produce an electric current (**I**), shown in the direction of the thumb in Figure 1.7. This current **I** produces a magnetic field **B** in the direction of the fingers in Figure 1.7. The rule that describes the direction of the current and its associated magnetic field is called the *right-hand rule*, because of the use of the right hand to describe it. This rule can be used in two ways. First, the thumb can point in the direction of the current, and the fingers represent the magnetic field (as shown). Alternatively, the thumb can point in the direction of the magnetic field, and the fingers will represent the direction of the current that produced it. The fact that the magnetic field encircles the current will be discussed in more detail in Section 1.5.

The magnetic field does not produce a force on a stationary charge (like the electric field does), but it does produce a force on any charge that is moving (in addition to that produced by the electric field). The force on a moving charge Q_{test} moving at a velocity **v** at a point P in space is illustrated in Figure 1.8(a). The force on the moving charge has a magnitude of $F = Bv\,Q_{test}$, where **B** is the *magnetic flux density*. The direction of the force is perpendicular to both **v** and **B**, as shown in Figure 1.8(b). The unit* of **B** is the tesla (T). Magnetic flux density is sometimes referred to as just *magnetic field.*[†]

Figure 1.9 shows vector plots of the **B** produced by a line current (an infinitely long current) and by a loop current. The **B** produced by the line current is strongest near the current, as indicated by closer spacing of the arrows. In each case the **B** lines encircle the current, which is a characteristic described in more detail in Section 1.5.

HOW MAGNETIC FIELDS ARE MEASURED

Magnetic fields are picked up using loops of wire, and in turn measuring the induced voltage across the ends of the wire (as discussed in the next section). The loop may be single or may be a coil of multiple loops, with the loop oriented so that the magnetic field lines pass through the loop. A typical configuration on a commercial magnetic field probe is shown in Figure 1.10. As with the electric field, three separate perpendicular loops can be used to pick up the three components of the magnetic field, as shown in Figure 1.11.

---

* A tesla is equivalent to an ampere-henry per square meter. The ampere is a unit of current. The henry is a unit of inductance.

† The related quantity **H** (see Section 1.7) is also often called *magnetic field*. The context is used to keep the meaning clear.

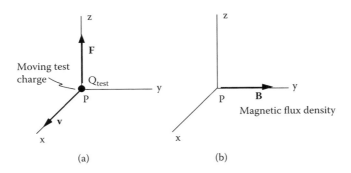

FIGURE 1.8
(a) Force **F** exerted by a magnetic field on a test charge having velocity **v** at a point P in space. **F** is perpendicular to **v**. (b) Magnetic flux density **B** defined at point P to account for **F**. **B** is perpendicular to both **v** and **F**.

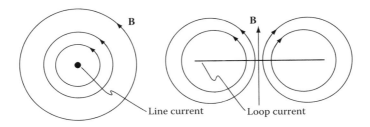

FIGURE 1.9
B fields produced by a line current and a loop current out of the page. The diagram shows just the edge of the loop current.

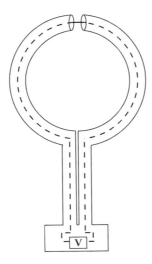

FIGURE 1.10
Loop antenna used for measuring magnetic fields. The antenna is made from two semirigid coaxial cables. (From Furse, C., et al., *Modern Antennas*, Wiley-Liss, Inc., a subsidiary of John Wiley & Sons, Inc., © Wiley-Liss 2007. With permission.)

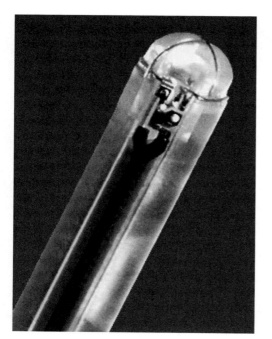

FIGURE 1.11
Magnetic field probes manufactured by SPEAG. Tip diameter is 6 mm. (From Schmid & Partner Engineering AG, Zurich. Reprinted with permission.)

1.4 Sources of Electric Fields (Maxwell's Equations)

Because **E** fields are defined to account for the forces exerted by charges on each other, the fundamental sources of **E** fields are electric charges. Specific information about how charges act as sources for **E** fields is given by Maxwell's equations, which are a fundamental set of equations that form the framework of all of classical electromagnetic field theory. Although we are minimizing the mathematical content of this book, we do state Maxwell's equations below because they are so fundamental and so famous in electromagnetics that we feel you should be introduced to them, even if you may not have a background in vector calculus and partial differential equations. We will just explain the qualitative meaning of these equations without giving the mathematical details.

Two of Maxwell's equations describe sources of **E**. One source is a time-varying **B** field, and the other is charge density ρ. Each source produces **E** fields with specific characteristics. For clarity, we describe these when each source is acting alone, but in general the **E** is produced by a combination of sources.

The first of Maxwell's equations that we discuss is Faraday's law:

$$\nabla \times \mathbf{E} = -\partial \mathbf{B}/\partial t \tag{1.3}$$

$\nabla \times \mathbf{E}$ is a mathematical expression called the *curl* of **E**, which means that the **E** produced will encircle the **B** that produced it. $\partial \mathbf{B}/\partial t$ is the time rate of change* of **B** (how fast **B** changes). This equation tells us that a time-varying magnetic field **B** creates an electric

---

* $\partial \mathbf{B}/\partial t$ is the time derivative of **B**. The symbol ∂ means change. So $\partial \mathbf{B}/\partial t$ means change in **B** ($\partial \mathbf{B}$) per change in t (∂t).

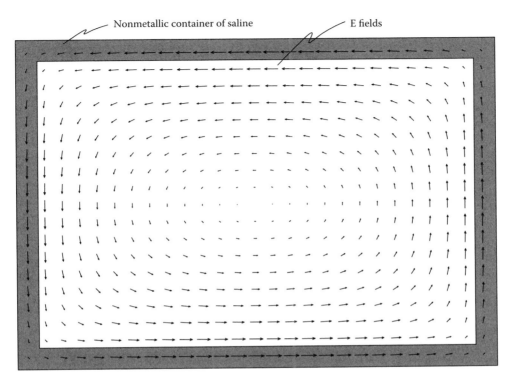

FIGURE 1.12
Calculated E fields at one instant of time for a two-dimensional model consisting of a 1 kHz **B** field (directed out of the paper) applied to a nonmetallic container of saline. The electric fields "curl" around the magnetic field.

field **E**. Generally speaking, the greater the time rate of change of **B**, the stronger **E** field it produces. This new **E** will also be time varying. There are many sources of time-varying magnetic field. Anything that uses typical commercial power (plugs into the wall) has 60 Hz fields.* This means that the fields vary sinusoidally (rise and fall) sixty times per second (see Section 1.9). Fields that vary sinusoidally with time are called *alternating current* (AC) fields. Sixty hertz is a relatively slow change in magnetic field, and therefore the electric field produced is quite small. Generally we approximate these fields as being constant with time. Fields that do not vary with time (such as those produced by a battery or permanent magnet) are called *direct current* (DC) fields. In the DC case, the magnetic field does not produce an electric field, and we say the fields are *decoupled*. Faster changes in magnetic fields are created in communication systems such as cellular telephones, which operate at 1,800 to 1,900 MHz.† These sources are also sinusoidal, and the time derivative of the magnetic field is on the order of 10^9 higher than for the 60 Hz commercial power case. Thus, the time-varying magnetic field generates a significant electric field. Other applications utilize fields that are pulsed, such as many imaging applications. For example, some types of microwave tomography use pulses containing frequencies from 300 to 3,000 MHz, and a new type of microwave breast imaging called confocal imaging uses pulses with frequencies up to 5,000 MHz. These sources are not sinusoidal. They are bursts of energy called *ultrawideband* (UWB) pulses, and they rise and fall very quickly (microseconds to

---

* Sixty hertz is the standard power frequency in the United States. Fifty hertz is used in Europe and Asia.
† 1 MHz = 10^6 Hz = 1 million Hz.

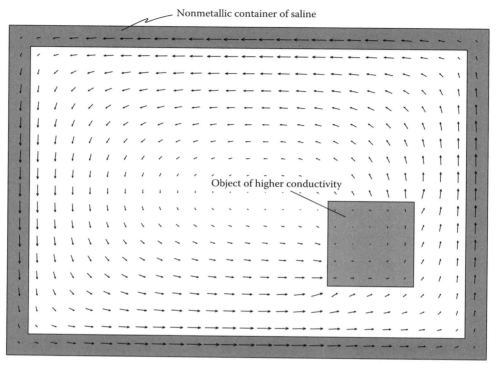

FIGURE 1.13

The same configuration as in Figure 1.12, but with an object of higher conductivity placed in the saline. The electric fields are smaller in the higher-conductivity object. The electric fields would also be smaller if the object had the same conductivity but higher permittivity than the saline.

nanoseconds). Thus, their time rate of change is very high, and a significant electric field is created from the time-varying magnetic field.

Figure 1.12 shows an example of the **E** fields in a nonmetallic container of saline produced by a changing **B** as calculated from a two-dimensional model.* The **E** field lines encircle (curl around) the changing **B**, which is directed out of the paper. Figure 1.13 shows the same configuration with an object added to the saline that has a higher conductivity (see Section 1.6) than the saline. Here again, the **E** field lines tend to encircle the changing **B**, but they are modified by the presence of the small object having higher conductivity. The higher conductivity of the small object causes the **E** fields inside the object to be weaker than those in the saline. The **E** field pattern in the small object can be thought of as consisting of two components: (1) the globally circulating **E** field of Figure 1.12 without the small object, and (2) an **E** field component circulating locally around the center of the small object. The resulting net pattern is a combination of the two, as shown in the magnified view of the object in Figure 1.14. On the left side and near the top of the object, the globally circulating **E** tends to cancel with the locally circulating **E**, while on the right side and near the bottom of the object, the two fields tend to add, producing a circulating pattern offset from the center of the object.

---

* A two-dimensional model is constant or equal in the third dimension, in and out of the page.

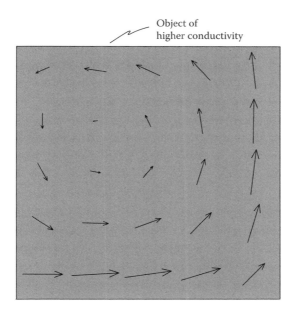

FIGURE 1.14
A magnified view of the **E** fields in the small object of higher conductivity of Figure 1.13. On the left side and near the top of the object, the globally circulating **E** tends to cancel with the locally circulating **E**, while on the right side and near the bottom of the object, the two fields tend to add, producing a circulating pattern offset from the center of the object.

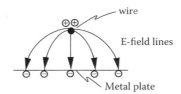

FIGURE 1.15
E field produced by positive charges on a wire and negative charges on a metal plate resulting from a potential difference applied between them.

A second of Maxwell's equations, Gauss' law, describes the **E** produced by charge density:

$$\nabla \bullet \mathbf{E} = \rho / \varepsilon \tag{1.4}$$

The expression $\nabla \bullet \mathbf{E}$ is called the divergence of **E**, which means an **E** field is created that starts at the source, ρ, which is the electric charge density in Coulombs per cubic meter (C/m³). ε is a parameter called *permittivity*, or *dielectric constant* (see Section 1.6), that just changes the magnitude of the electric field but does not create it or change its direction. Equation 1.4 means that electric charge creates **E**, and that the **E** lines begin and end on charges.

Figure 1.15 shows an example of the **E** fields produced by charges. A potential difference applied between a long wire and a metal plate produces positive charges on the wire and negative charges on the plate. These charges produce the kind of **E** field lines shown.

1.5 Sources of Magnetic Fields (Maxwell's Equations)

Another two of Maxwell's equations describe sources of **B**. Ampere's law states that

$$\nabla \times \mathbf{B} = \mu(\mathbf{J} + \varepsilon \, \partial \mathbf{E}/\partial t) \tag{1.5}$$

where μ is a constant called *permeability* (Section 1.6) that affects the field magnitude but does not produce it or change its direction. As with Faraday's law, $\partial \mathbf{E}/\partial t$ represents the rate of change of the electric field. Ampere's law shows that current density **J** (A/m^2) and a time-varying electric field $\partial \mathbf{E}/\partial t$ are both sources of **B**, and that the **B** field lines produced by these two sources encircle (curl around) **J** and $\partial \mathbf{E}/\partial t$. The magnetic field produced by the electric field will always be time varying (AC). The magnetic field produced by **J** may be either AC or DC depending on **J**.

And finally, the last of Maxwell's equations, Gauss' law for magnetism, is

$$\nabla \bullet \mathbf{E} = 0 \tag{1.6}$$

This equation states that the divergence of **B** is always zero, which means that there are no magnetic charges analogous to electric charges, and that **B** field lines always occur in closed loops since they do not begin and end on charges, as do **E** fields.

Figure 1.9 shows examples of how current density **J** produces **B** fields, and how the **B** field lines encircle the current. At low frequencies, the time-changing **E** field is usually a weak source compared to **J**, and so typical low-frequency systems do not involve significant **B** fields produced by $\partial \mathbf{E}/\partial t$. We postpone discussion of examples showing how $\partial \mathbf{E}/\partial t$ produces **B** until Chapter 3.

INDUCTIVE TELEMETRY FOR COMMUNICATION WITH MEDICAL IMPLANTS

Implantable medical devices such as cardiac pacemakers and defibrillators (as shown in Figure 1.16), neural recording and stimulation devices, and cochlear and retinal implants require methods to recharge their batteries and transmit data both to and from the device. Inductive coupling is the most common method of doing this today. Inductive coupling works by utilizing an alternating current I_1 in a loop of wire, as shown in Figure 1.17. The magnetic field **B** that is caused by this current (see the right-hand rule in Section 1.3) passes through a second (parallel) loop, where it generates a second current I_2. If one of the loops is on the inside of the body and the other on the outside, the magnetic field will pass relatively unchanged through the body to the second loop. The current generated on the second loop can be used to recharge a battery or send data to an electrical device inside the body. Inductive coupling works best if the two loops are very close together and perfectly aligned parallel to each other. Otherwise, the magnetic field spreads out, and not all of it is picked up by the second loop. Using more loops (coils) will increase the amount of coupling (how much current I_2 is generated from current I_1).

Inductively coupled applications are usually limited to transcutaneous links rather than transmission through larger, more lossy regions of the *(continued on next page)* body. Typically, inductively coupled coils are wound around a ferrite core to improve the amount of magnetic field that can be transmitted from one coil through the skin to the other coil. Frequencies are often lower than 50 MHz to ensure that the presence of the human body (skin) does not significantly obstruct the coupling between the coils.

Most inductive telemetry links are used for subcutaneous applications due to power restrictions for implanted devices. Data rates are generally low, and size/weight and biocompatibility issues plague these devices. However, recent advances continue to reduce the power requirements and provide more biocompatible designs. For example, the Utah Electrode Array (Figure 1.18) has an array of one hundred tiny silicon electrodes that each pick up the nerve signal from a single neuron. A computer chip is integrated into the top of the electrode array in order to receive and process the signals from the electrodes. In order to receive external power and to upload and download data, a pickup coil is printed on a ceramic substrate and integrated with the implanted neural electrode array, as shown. The implanted coil is energized by an external inductive programmer/reader that powers the implanted circuitry while transferring telemetry data.

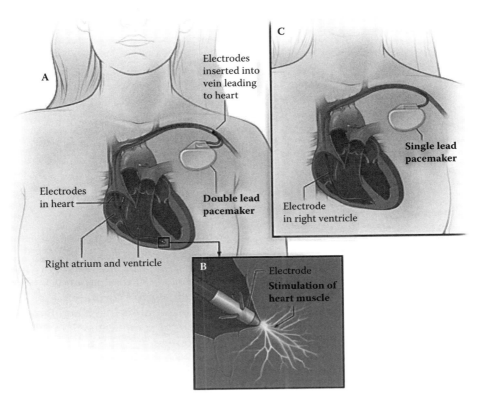

FIGURE 1.16
Example of an implanted pacemaker. (a) Double-chamber (double-lead) pacemaker. (b) Electrode electrically stimulating heart muscle. (c) Single-chamber (single-lead) pacemaker. (From the National Heart, Lung, and Blood Institute as a part of the National Institutes of Health and the U.S. Department of Health and Human Services.)

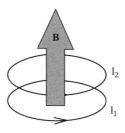

FIGURE 1.17
An alternating current I_1 in one loop will produce a magnetic field B that, upon passing through the second loop, will produce a current I_2 in that loop.

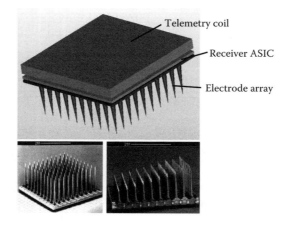

FIGURE 1.18
Utah Electrode Array packaged with a custom ASIC and printed receiver coil. (Top figure: From Florian Solzabacher. With permission. Bottom figures: From Guillory, K., and Normann, R. A., *J. Neurosci. Methods*, 91, 21–29, 1999. With permission.)

1.6 Electric and Magnetic Field Interactions with Materials

One of the more important aspects of bioelectromagnetics is how electromagnetic fields interact with materials, for example, how **E** and **B** fields affect the human body and how the body affects the fields. Because **E** and **B** were defined to account for forces among charges, the fundamental interaction of **E** and **B** with materials is that **E** and **B** exert forces on the charges in the materials. The interaction is even more complicated than that, though, because the charges in materials also act in turn as sources of **E** and **B**. The *applied* fields, as they are often called, are produced by source charges external to a given material in the absence of the material. The *internal* fields in the body are the combination of the applied fields and the fields produced by the charges inside material. The *scattered fields* are fields external to the object, produced by charges inside the object. Usually in an electrically neutral object, the algebraic sum of the positive and negative charges inside the object is zero, and the positive and negative charges are microscopically so close together that the fields

they produce cancel on a macroscopic scale. The
applied fields, however, exert forces on the internal charges, which cause them to separate so that
the macroscopic fields they produce no longer cancel. These fields combine with the original applied
fields to produce a new internal field, which further
affects the internal charges. This process continues
until an equilibrium is reached, resulting in some
net internal field.

Electric dipole resulting
from charge separation

FIGURE 1.19
Illustration of how an **E** causes charge separation, which results in an electric dipole,
the combination of a positive and a negative
charge separated by a very small distance.

In most cases, accounting for the interaction
with charges in a material on a microscopic scale is
impossible in practice. The interaction is therefore
described macroscopically in terms of three effects
of fields on the charges in the material: induced dipole polarization, alignment of already
existing electric dipoles, and movement of *free* charges. Figure 1.19 illustrates the concept
of induced dipoles. Before the **E** is applied, the positive and negative charges are so close
together that the macroscopic fields they produce cancel each other. When an **E** field is
applied, the positive charge moves in one direction and the negative charge in the opposite direction, resulting in a slight separation of charge. The combination of a positive and
a negative charge separated by a very small distance is called an *electric dipole*. These are
bound charges, because they are held in place by molecular bonds and are not free to move
to another molecule. The creation of electric dipoles by this separation of charge is called
induced polarization.

In some materials, such as hydrogen-based biological materials, electric dipoles already
exist, even in the absence of an applied **E** field. These permanent dipoles are randomly
oriented, so that the net fields they produce are zero.
When an **E** field is applied, the permanent dipoles
partially align with the applied **E**, as illustrated in
Figure 1.20. The applied **E** exerts a force on the positive charge of the dipole in one direction and on
the negative charge in the opposite direction, causing the dipole to rotate slightly, and thus partially
align with the applied **E**. This partial alignment of
the permanent dipoles reduces the randomization
so that the net **E** field produced by the collection of
dipoles is no longer zero.*

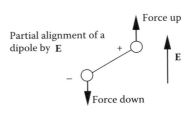

Partial alignment of a
dipole by **E**

FIGURE 1.20
Illustration of partial alignment of a permanent electric dipole by an applied **E** field.

The third effect of applied **E** fields on material
charges is illustrated in Figure 1.21. Some charges (electrons and ions) in materials are free
in the sense that they are loosely bound, and can move between molecules in response
to an applied **E** field. These charges move a short distance, collide with other particles,
and then move in a different direction, resulting in some macroscopic average velocity in
the direction of the applied **E** field. The movement of these free charges constitutes a current, which is called *conduction current*. Metals and high-water-content tissues have more

---

* Because charges are effectively repositioned inside the material by either induced polarization or alignment
  of permanent dipoles, current appears to be produced; this type of apparent current in combination with the
  rate of change of electric field is called *displacement current* (see Section 2.4.2), and it plays a key role in capacitors and in the propagation of EM waves. Displacement current is enhanced by movement of bound charges,
  while conduction current is caused by movement of free charges.

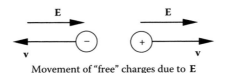

Movement of "free" charges due to **E**

FIGURE 1.21
"Free" charges in materials acquire a velocity in response to an applied **E** field. Positive charges move in the same direction as **E**, and negative charges move in a direction opposite to the direction of **E**.

free charges than insulating materials (such as glass, plastic, bone, or fat). The conduction currents that are carried by metal wires are what we typically associate with electricity. Conduction currents also cause heating and power loss in a material. More information on conduction current is found in Section 1.14.

Similarly to how **E** causes partial alignment of permanent electric dipoles in materials, **B** causes partial alignment of permanent magnetic dipoles in materials (but there is no effect of **B** analogous to the separation of electric charge by an applied **E** field). The alignment of magnetic dipoles becomes very important during *magnetic resonance imaging* (MRI) applications, as described in Section 6.4.

Because the interactions of **E** and **B** with materials are too complex to keep track of in terms of individual charges, three parameters are defined to account for these interactions on a macroscopic scale. Induced polarization and alignment of permanent electric dipoles is accounted for by *permittivity*, also called *dielectric constant*, which describes how much induced polarization and partial alignment of permanent electric dipoles occurs for a given applied **E**. Conduction current is accounted for by *conductivity*, which describes how much conduction current density a given applied **E** will produce. Alignment of permanent magnetic dipoles is accounted for by *permeability*, which describes how much partial alignment of permanent magnetic dipoles occurs for a given applied **B**.

Permittivity is often represented by the Greek letter epsilon (ε); its units* are farads per meter (F/m). The permittivity of free space (no charges present) is called ε_0 and in the International System of Units (SI), $\varepsilon_0 = 8.854 \times 10^{-12}$ F/m. *Relative permittivity* is defined as $\varepsilon_r = \varepsilon/\varepsilon_0$; it is the permittivity relative to that of free space, and is unitless. Conductivity[†] is often represented by the Greek letter sigma (σ); its units are siemens per meter (S/m), which is the same as 1/ohm-m. Permeability is usually represented by the Greek letter mu (μ); its units[‡] are henrys per meter (H/m). The permeability of free space is $\mu_0 = 4\pi \times 10^{-7}$ H/m, and *relative permeability* is defined as $\mu_r = \mu/\mu_0$; it is unitless. For most applications, the human body is so weakly magnetic that we can assume $\mu = \mu_0$, so $\mu_r = 1$. Appendix A discusses the electrical properties of specific human tissues in more detail.

---

* The farad is the unit of capacitance, which represents storage of charge.
† In practice, the parameter for conductivity normally includes two components. The first (σ_c) represents conductivity due to free charge movement, leading to the classical definition of conduction current $J_c = \sigma_c \mathbf{E}$. In addition, the displacement current caused by motion of bound charges may have a portion that is in phase with the electric field. This in-phase term accounts for the energy loss associated with the motion of bound charges and can be represented by another component of conductivity, $\sigma_d = \omega \varepsilon_0 \varepsilon''$. Together they form the *effective* conductivity, $\sigma_{eff} = \sigma_c + \sigma_d$. (Effective conductivity is often simply called conductivity, and many books just use the variable σ to represent it.) The effective conductivity can be used to determine the total loss of an electromagnetic wave as it passes through a material and the effective conduction current, $J_{c,eff} = \sigma_{eff} \mathbf{E}$. More detail is given in Section 1.14.
‡ The henry is the unit of inductance, which represents resistance to time-varying change of current.

1.7 Other Electromagnetic Field Definitions

Two other definitions are used in electromagnetic field theory. One of them is *magnetic field strength* or *magnetic field intensity*, defined* as

$$\mathbf{H} = \mathbf{B}/\mu \ (A/m) \tag{1.7}$$

which has units of amperes per meter (A/m). As discussed in Chapter 3, **H** is often more convenient to use than **B** in describing EM wave interactions. In practice, both **B** and **H** are often referred to simply as magnetic fields.

The other definition is *electric flux density* or *electric displacement*, defined as

$$\mathbf{D} = \varepsilon \, \mathbf{E} \ (C/m^2) \tag{1.8}$$

which has units of coulombs per square meter (C/m²). Sometimes using **D** is more convenient than **E** in EM field theory, but in this book we will mostly be using **E**.

1.8 Waveforms Used in Electromagnetics

The shape of the wave as a function of time is called the *waveform*. The simplest type of electrical waveform is constant with time. This is called *direct current* or *DC* (also called *static* fields). A battery is a good example of a DC power source that can be connected to a load such as a lightbulb, as illustrated in Figure 1.22. The voltage on the battery (V volts) is defined with respect to positive and negative terminals, just as they are labeled on a battery. The current (I amps) is defined as flowing from the positive to the negative terminal through the resistor (R ohms). A lightbulb is a resistor that converts the cur-

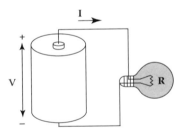

FIGURE 1.22
A simple circuit illustrating the polarization (+/−) of a DC voltage source (battery) and the current. The lightbulb filament acts as the resistor R in this circuit.

rent in its filament to light (and heat). Ohm's law states that V = I R. The power that is dissipated in the resistor is given by P = V²/R = I² R (watts). One of the most important aspects of DC fields for bioelectromagnetic applications is that the electric and magnetic fields do not generate each other. They are therefore decoupled (as indicated earlier in Section 1.4), and each one can be analyzed separately. Also, a DC signal requires a conductive path in order to propagate, and the current seeks the path of least resistance. In our lightbulb example, the DC current follows the wire. If the wire is broken or disconnected (such as by a switch), the current no longer flows. Low-frequency waves (typically less than 10 MHz)

---

* When the permeability μ or permittivity ε is a complex quantity, as described in Section 1.14, Equations 1.7 and 1.8 still apply if μ or ε is replaced by its complex counterpart.

are often approximated as being static, and are called *quasi-static*. They are then treated much like DC fields.

Many applications in electromagnetics use waves that are single-frequency sine waves as described in the next section. These waves are often called *alternating current* or *AC* waves, because the current alternates from positive to negative and back again. Cell phones, television and radio stations (anything with an antenna), microwave ovens, and light fall into this category. A major characteristic of AC waves is that the electric and magnetic fields can generate each other and are therefore coupled and cannot be analyzed separately. This also means that the signals are not required to stay on a wire, and can therefore propagate through the air or other materials. These signals generally choose the path of least inductance rather than the path of least resistance. Another important consideration of this type of field is that a single wire does not necessarily ground the device. Typically multiple connections in parallel or a metal plate or grid are used to provide a good AC ground. This can become important in electrotherapy applications.

Other applications use electrical signatures that are more complex and are a combination of DC and multiple-frequency AC fields. For instance, pulsed systems are used for electrotherapy for pain control, bone and tissue healing, and muscle stimulation. These systems use the many different types of pulses shown in Figure 1.23. Even though the symmetrical biphasic pulse may look visually like an AC sine wave, described in the next section, it is not. The fact that the pulse starts and stops (as opposed to continuing on forever, as the sinusoid does) is very important. A pulse can be represented mathematically as a sum of sine waves of different frequencies (a Fourier series). The frequencies of the individual sine waves required to represent the pulse are called the frequency content of the pulse. A continuous sine wave has only a single frequency, and sums of sine waves have only the frequencies in the waves that are summed up. Pulses typically have a much broader range of frequencies (that is, they consist of the sum of many individual sine waves of different frequencies), even if they are simply a biphasic sinusoid that stops and starts.

It is important to note here that electromagnetic waves for biological systems are almost always linear with respect to frequency. This means that the result from a combination of frequencies would be the same whether we used all of the frequencies simultaneously or whether we used them sequentially and then added up the effects from each frequency

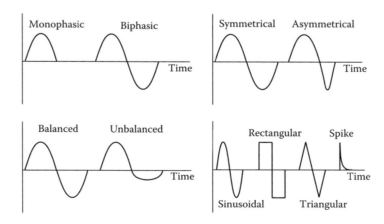

FIGURE 1.23
Several types of pulses used in electrophysiological applications.

individually. This is very fortunate, because it means that the sinusoidal analysis described throughout this book can be used for virtually any application, even pulsed applications.

There are some simple ways to determine the relative frequency content of different pulses. First, the average value of the pulse over time gives the DC (zero frequency) term. If the average over time is zero, the DC term is also zero. Symmetric, balanced biphasic pulses have zero DC frequency content, and anything else has nonzero DC content. This means that any pulse with a DC component will exhibit at least some of the characteristics seen in Chapter 2 (which deals with low-frequency or quasi-static signals). Pulses with faster rise times (the rectangular and spike pulses, particularly) will have the highest frequency content. High frequencies are contained in fast changes. In fact, if the rectangular pulse actually rose instantaneously, it would have infinite frequency content. Of course, this is impossible in a real system, and rectangular pulses have a small but finite rise time. It might seem as if the sinusoidal pulse should have only one frequency, like the AC sine wave. This is not the case, however, because of the turn-on and turn-off points. The sinusoidal pulse is actually a sine wave multiplied by a monophasic rectangular pulse and has the frequency content of both. A good rule of thumb for determining the frequency content of a pulse is to ignore any flat sections (like the top of the rectangular pulse), and then the narrower the pulse is in time, the higher the frequencies it contains.

1.9 Sinusoidal EM Functions

Sinusoids, or *sinusoidal wave functions*, are widely used to describe behaviors in physical systems, including electromagnetic systems. Figure 1.24 shows a sine wave, f(x) = sin(x), and a cosine wave, g(x) = cos(x). x is called the *independent variable* and f and g are *dependent variables*. These functions are both called sinusoids, because they are described by the trigonometric functions sin(x) and cos(x). Values of the functions sin(x) and cos(x) for various values of x can be found in mathematical books and tables and from engineering and scientific calculators.

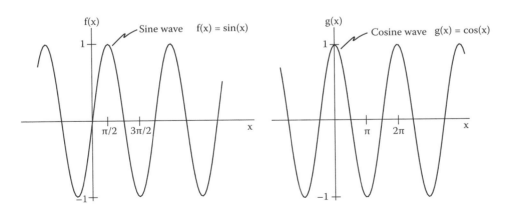

FIGURE 1.24
Sinusoidal functions of x: a sine wave and a cosine wave. Although not shown, these waves theoretically extend to infinity on both ends of the horizontal axis.

In electromagnetics, the typical independent variables are space (z) and time (t). Here we will use time as the independent variable in describing the properties of sinusoids. When time is the independent variable, the function is said to be in the *time domain*. Any sinusoidal function of time can be written in the general form

$$g(t) = A \cos(\omega t - \phi) \qquad (1.9)$$

where A is called the *amplitude*, the *peak value*, or the *maximum value*; ω is the *radian* or *angular frequency*; and ϕ is the *phase angle*. Figure 1.25 shows g(t) plotted both as a function of t and as a function of ωt. The *period* T is defined as the time between any two corresponding similar points on the waveform, such as between the two peaks in Figure 1.25(a). The frequency f is defined as

$$f = 1/T \ (\text{Hz} = 1/s) \qquad (1.10)$$

with units of hertz (Hz). The radian frequency (or angular frequency) ω is related to f by

$$\omega = 2\pi f \ (\text{rad/s}) \qquad (1.11)$$

with units of radians per second (rad/s). You have probably noticed that $g(t) = A \cos(\omega t - \phi)$ is just $A \cos(\omega t)$ shifted to the right. Figure 1.25(b) shows that ϕ is the angle in radians by which $A \cos(\omega t)$ is shifted to the right to produce $A \cos(\omega t - \phi)$, when plotted against ωt. Note that when $\phi = \pi/2$ radians, $A \cos(\omega t - \pi/2)$ is exactly the same as $A \sin(\omega t)$. This illustrates the fact that any sinusoidal function of time can be written in the form of Equation 1.9.

Two sinusoids expressed in the form of Equation 1.9 are said to be *in phase* if their phase angles are equal, which means they line up in time. They are said to be *out of phase* when their phase angles are not equal, in which case they will not line up in time. Figure 1.26

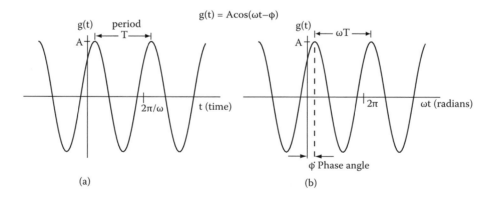

FIGURE 1.25
General form of a sinusoidal function or wave. (a) is plotted as a function of time t, and (b) is plotted as a function of phase (ωt). The period of the wave T is the distance between two identical points on the wave, as shown in (a) between the peaks. The phase shift is ϕ, as shown in (b); this is equivalent to a time delay in (a). If the magnitude A was increased, the wave would get taller. If the frequency ω was increased, the peaks would be closer together and the period T would be smaller. If the phase shift ϕ was increased, the wave would move farther to the right.

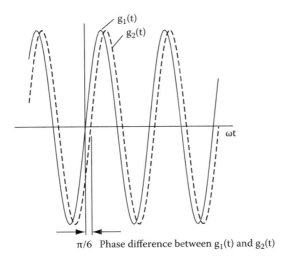

π/6 Phase difference between $g_1(t)$ and $g_2(t)$

FIGURE 1.26
Two sinusoidal functions out of phase by π/6 radians. Function $g_2(t)$ lags $g_1(t)$ (i.e., it arrives at a later time). Remember, 360° = 2π radians, so π/6 is equivalent to 30°.

shows two functions, $g_1(t)$ and $g_2(t)$, that are out of phase by π/6 radians. Phase angles and differences in phase are often specified in degrees, done by converting the angles in units of radians to angles in units of degrees, although it is not strictly correct to do so because ωt has units of radians and φ and ωt must have the same units. (To convert from radians to degrees, multiply the radians by 180/π. To convert from degrees to radians, multiply the degrees by π/180.) Thus, $g_1(t)$ and $g_2(t)$ are said to be out of phase by π/6 radians, or 30°.

1.10 Root Mean Square or Effective Values

In many instances, it is convenient to describe time-varying fields in terms of their *root mean square* (rms) values. Of particular importance is the use of rms values in describing average power, which most EM equipment measures. The relationship between average power and rms values is illustrated in Figure 1.27. The instantaneous (not average) power transferred to tissue by a time-varying **E** field is proportional to E^2 at any instant of time (see Section 1.16). For example, if **E** is a sinusoidal function of time, the instantaneous power transferred will be proportional to the square of a sine wave, as shown in Figure 1.27. This instantaneous power fluctuates from zero to some maximum value, which is proportional to the peak value of E^2.

The *average* value of the power, which is usually of prime importance, is proportional to the average value of E^2, also shown in Figure 1.27. In this illustration, the peak value of E is 1.5, the peak value of E^2 is 2.25, and the average value of E^2 is 1/2 of 2.25, which is 1.125. But the average value of E^2 can also be written as $(1.5/\sqrt{2})^2$. Therefore, in this illustration, the quantity $(1.5/\sqrt{2})$ is the rms value of the function E. In general, the average value of the *square* of a sinusoidal function is equal to $[\text{peak value}/\sqrt{2}]^2$. Thus, the rms value of any sine wave is its peak value divided by $\sqrt{2}$. For example, the rms value of **E** is

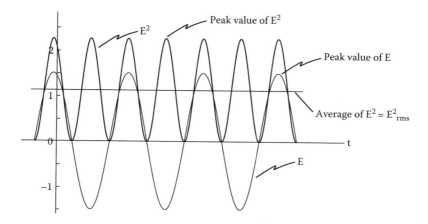

FIGURE 1.27
E is a sinusoidal function of t. Also shown is E^2. The average value of E^2 is one-half the peak value of E^2, also equal to E^2_{rms}.

$$E_{rms} = E / \sqrt{2} \qquad\qquad (1.12)$$

where E is the peak value of the **E** field. The rms value is also called the *effective value*, because it has the same effect in producing average power as a steady function of the same value that does not vary with time.

   In general, as given by its name, the rms value of a function is defined as the square root of the mean of the square of the function. Thus, to find the rms value of a given function, first square it, then find the mean (average) of the squared function, then take the square root of that. For sinusoidal functions, this procedure always gives an rms value that is equal to the peak value divided by $\sqrt{2}$. As an example of finding the rms value of a non-sinusoidal function, let us calculate the rms value of the function f shown in Figure 1.28, which is a periodic function of t. First we square f as shown in the bottom graph. Then we find the average of f^2 by finding the area between f^2 and the t axis, which is shown shaded for one period of f in the figure. The area is $9 \times 1 + 4 \times 3 = 21$. The average value of f^2 is this area divided by the period (which is 4). Thus, the average of f^2 is 21/4. The rms value of f is the square root of the average of f^2, which is $\sqrt{21}/2 = 2.29$.

1.11 Wave Properties in Lossless Materials

For many physical configurations, the solutions to Maxwell's equations are most conveniently formulated in terms of *propagating* sinusoidal wave functions, or *waves*. Because an understanding of the properties of waves is essential for much of the remainder of the book, we review those properties here. This section covers lossless materials, where $\sigma = 0$. Lossless materials are sometimes also called *perfect dielectrics*. No power is lost or deposited in a lossless material, and electromagnetic waves will not cause it to heat up. Glass and most plastics are lossless or near-lossless materials. Lossless materials make good electrical insulators, because they contain no free charges. Waves in lossy materials

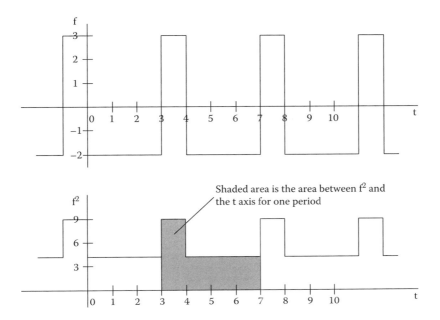

FIGURE 1.28
Illustration for calculating the rms value of the function f. The top graph shows f as a function of t, and the bottom one shows f^2 as a function of t.

are covered in Section 1.14. Figure 1.29 illustrates the concepts of propagating waves. As an example, the figure shows the magnitude of the electric field **E** as a function of distance z at two different instants of time, t_1 and t_2. These plots can be thought of as snapshots of a wave that is propagating to the right. The waveform shown is called a *sinusoidal wave* (see Section 1.9) because it is described by either the trigonometric function $\sin(\omega t - \beta z - \phi)$ or $\cos(\omega t - \beta z - \chi)$. ω is the radian frequency in radians per second (rad/s), β is the *propagation constant* in radians per meter (rad/m), and ϕ and χ are the phase angles. The *wavelength* λ is defined as the distance in meters at one instant of time between any two corresponding

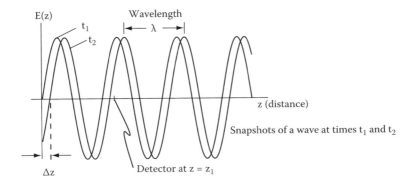

FIGURE 1.29
Illustration of a propagating wave. The magnitude of **E** is shown as a function of distance z at two different instants of time. The wave extends infinitely far in both the positive and negative z directions, though it is shown only over a limited range of z.

points on the wave. In the figure, the two corresponding points are neighboring peaks (maxima) of the wave.

The magnitude of the *phase velocity* v_p of the wave is defined as

$$v_p = \Delta z/\Delta t \ (\text{m/s}) \tag{1.13}$$

where $\Delta t = t_2 - t_1$ is a small difference in time, and Δz is a small distance that the wave travels in the time Δt. The phase velocity, as its name implies, describes how fast the wave is moving and is often also called the *velocity of propagation*. The relationship between ω, β, and v_p is

$$\beta = \omega/v_p \ (\text{rad/m}) \tag{1.14}$$

If a detector that registers the magnitude of **E** as a function of time were placed at a point z_1 on the z axis in Figure 1.29, the output of the detector as a function of time would be the sinusoidal function of time shown in Figure 1.30, with period T, frequency f, and radian frequency ω, as defined in Section 1.9. The wavelength, phase velocity, and frequency are related by

$$\lambda = v_p/f \ (\text{m}) \tag{1.15}$$

The phase velocity of a wave in free space is often designated by c or c_o. For planewaves (a specific kind of propagating wave that is described in Section 3.2.2) the phase velocity in free space is $c = 3 \times 10^8$ m/s. This is also the *speed of light* in a vacuum, since light is one of the waves that travels at this velocity. Combining Equations 1.11, 1.14, and 1.15 gives the relationship between the propagation constant β and the wavelength λ:

$$\beta = 2\pi/\lambda \ (\text{rad/m}) \tag{1.16}$$

Because sinusoidal functions are so prevalent in descriptions of electromagnetic fields, the characteristics of EM fields are often described in terms of the *frequency spectrum* or frequency range. Figure 1.31 shows a simplified representation of the electromagnetic frequency spectrum with both the frequency f and the wavelength λ in free space indicated.

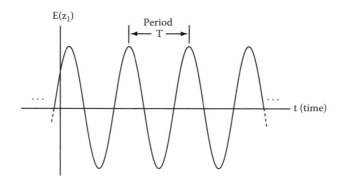

FIGURE 1.30
The output of a detector placed at position $z = z_1$ in Figure 1.29. The waveform extends infinitely far in both the positive and negative t directions, though it is shown only for a limited range of t.

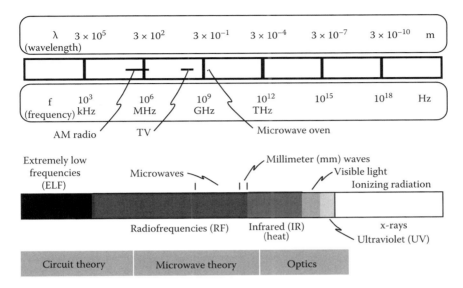

FIGURE 1.31
Simplified illustration of the extent of the electromagnetic frequency spectrum and how the characterizations of Table 1.1 fit into the frequency spectrum for typical systems.

Some of the familiar frequency bands, such as the AM radio broadcast bands and the television broadcast bands, are indicated in the figure. Maxwell's equations are valid over this whole range. It is astonishing, indeed, that one set of equations could be valid over more than 18 orders of magnitude in some parameter. Of course, even though Maxwell's equations apply over this whole range, and even though they are deceptively simple in form, they are certainly not easy to solve. No general solution that applies over the whole range is available. Instead, special techniques are used to find solutions for specific classes of problems in various frequency ranges. A general description of these techniques is given in Section 1.18, and more detail can be found in Chapter 5.

1.12 Boundary Conditions for Lossless Materials

Boundary conditions are relationships between EM fields that must be satisfied at the interface between two different materials, as required by Maxwell's equations. Because these boundary conditions are useful in interpreting and explaining characteristic behaviors of EM field interactions with biological systems, we discuss them here.

Figure 1.32 illustrates the boundary conditions on the **E** field. Because **E** is a vector, it can be resolved into two components, one parallel (tangential) to the boundary and one perpendicular (normal) to the boundary. Resolving vectors into components is explained in Figure 1.33, which illustrates how the vector **A** can be resolved into different pairs of components that are perpendicular (normal) to each other. Vectors are added graphically tail to head. Thus, the tail of $\mathbf{A}_2$ is placed at the head of $\mathbf{A}_1$, and then the sum of $\mathbf{A}_1$ and $\mathbf{A}_2$ is a vector from the tail of $\mathbf{A}_1$ to the head of $\mathbf{A}_2$, which in Figure 1.33(a) is the vector **A**. $\mathbf{A}_1$ and $\mathbf{A}_2$ are normal to each other. Figure 1.33(b) shows how two other vectors, $\mathbf{A}_3$ and $\mathbf{A}_4$,

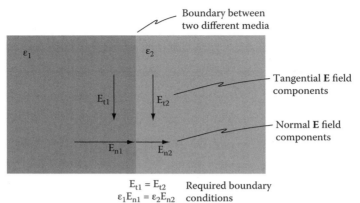

FIGURE 1.32
Illustration of conditions on the **E** field required by Maxwell's equations to be satisfied at the interface (boundary) between two media with different permittivities ε.

add up to the same vector **A**. $\mathbf{A}_3$ and $\mathbf{A}_4$ are also normal to each other. Thus, we say that **A** can be *resolved* into the components $\mathbf{A}_1$ and $\mathbf{A}_2$. **A** can also be resolved into components $\mathbf{A}_3$ and $\mathbf{A}_4$, as well as into many other components. Figure 1.33(c) shows how the vector **B** can be resolved into two components $\mathbf{B}_t$ and $\mathbf{B}_n$, where $\mathbf{B}_t$ is the component that is tangential (parallel) to a given boundary line and $\mathbf{B}_n$ is the component that is normal (perpendicular) to the boundary line.

Maxwell's equations require that the normal components of the **E** field at a charge-free boundary satisfy this equation:

$$\varepsilon_1 E_{n1} = \varepsilon_2 E_{n2} \qquad (1.17)$$

where E_{n1} is the normal component of the **E** field in medium 1 at the boundary, E_{n2} is the normal component of the **E** field in medium 2 at the boundary, ε_1 is the permittivity of

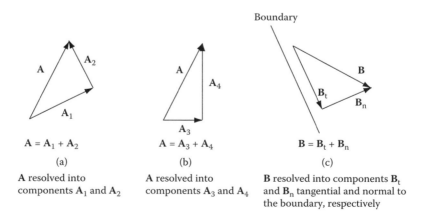

FIGURE 1.33
Resolving vectors into components. (a) **A** is resolved into components $\mathbf{A}_1$ and $\mathbf{A}_2$. (b) **A** is resolved into components $\mathbf{A}_3$ and $\mathbf{A}_4$. (c) **B** is resolved into components $\mathbf{B}_t$ and $\mathbf{B}_n$.

THE DIFFICULTY IN PROPAGATING THROUGH
SKIN-FAT-MUSCLE LAYERS

One of the great strengths of electromagnetic fields is that they can be transmitted, to one degree or another, from the exterior to the interior of the body. Over most of the regions of the body, this means transmitting from air ($\varepsilon_r = 1$), through skin ($\varepsilon_r = 51$ at 433 MHz, according to Appendix A), then fat ($\varepsilon_r = 5$), then muscle ($\varepsilon_r = 64$) or muscle-like material. Boundary conditions can give us a rough understanding of what happens in these layers of tissue. Suppose an electric field has 1 V/m components in both the tangential and normal directions. The tangential component will pass through all of the layers of tissue into the muscle tissue without changing, as indicated by Equation 1.18. (Here we are neglecting some major factors. The field would be attenuated by the lossy tissues, particularly muscle, and reflections would occur within the layers. But for a first-order approximation at low frequencies, we can assume that the tangential field passes through the layers unperturbed.)

The normal field, on the other hand, is definitely changed by the layers of tissue, as given by Equation 1.17. In the skin layer, the 1 V/m field becomes $(1 \text{ V/m})(\varepsilon_{r\,air}/\varepsilon_{r\,skin}) = 1/51$ V/m. The field in the skin then passes into the fat layer with a magnitude of $(1/51 \text{ V/m})(\varepsilon_{r\,skin}/\varepsilon_{r\,fat}) = 1/5$ V/m. In turn, the field in the fat passes into the muscle layer with a magnitude of $(1/5 \text{ V/m})(\varepsilon_{r\,fat}/\varepsilon_{r\,muscle}) = 1/64$ V/m. Note that the field in the fat is roughly ten times larger than the fields in either the skin or the muscle, while we generally want the largest field deep within the body in the muscle or muscle-like region. The boundary conditions show that the fat layer is likely to have a much larger field. This problem is exacerbated by the multiple reflections that occur in these layers at higher frequencies. A standing wave (see Section 3.3.1) can be set up in the fat layer, which can create even higher field concentrations. The large fields in the fat can create very painful subdermal electrical burns that are difficult to treat, and must be avoided. A common way of mitigating this effect is to place a bolus (pillow) of water between the electromagnetic source and the body. The water can be cooled to reduce the heat at and near the body surface. By propagating the field into the water first, the concentration of power at the surface of the body can be minimized, and more uniform field distributions result.

medium 1, and ε_2 is the permittivity of medium 2. Note that Equation 1.17 is required to be true only at the interface, or boundary, between the two media. The **E** field may change significantly as a function of position in the medium, and Equation 1.17 is not required to be satisfied at any points other than those on the boundary.

The boundary condition for the tangential components of **E** at the boundary is

$$E_{t1} = E_{t2} \tag{1.18}$$

where E_{t1} and E_{t2} are the tangential components of **E** at the boundary in mediums 1 and 2, respectively. One of the important impacts of this equation is that on the surface of metal, the tangential electric field is always zero. This is because there is no electric field inside an ideal metal (either tangential or normal), and therefore the tangential field on the outside surface is also zero. This has implications in many aspects of electromagnetics. For instance, in the design of probes to measure electric fields described in Section 1.2, the metal arms of the measuring antenna will have zero tangential electric field. The tangential electric field that is very near but not directly in contact with the arms of the antenna will be detected, however, through a process where the associated magnetic field creates a measurable current on the antenna. But merely placing the metal antenna into the field forces the tangential electric field on the antenna to zero, and slightly impacts the fields nearby. If the antenna were very large, the presence of the antenna would significantly perturb all of the fields the antenna is trying to measure, creating erroneous readings. Note that special care must also be taken for all of the wires that connect these antennas to the outside measurement equipment, because the metal wires could also perturb the fields. To avoid field perturbation, these are often made from high-resistivity plastic, not metal at all!

As an example of the boundary conditions, let us suppose that medium 1 in Figure 1.32 is air, and medium 2 is muscle tissue. At low frequencies, $\varepsilon_2/\varepsilon_1 \approx 10^6$. From Equation 1.17, this means that $E_{n2}/E_{n1} \approx 10^{-6}$. The normal E field in the muscle at the boundary would therefore be much smaller than the normal E field in the air at the boundary. On the other hand, from Equation 1.18, the tangential components of E in the air and in the muscle are equal at the boundary. Because the E in the air is a combination of incident and scattered fields (Section 1.6), it is difficult to draw definitive conclusions about the relative magnitudes of the internal E field components compared to the incident E field components from boundary conditions alone. But it turns out that when the incident E is mostly normal to biological tissue at low frequencies, the internal E is smaller than when the incident E is mostly parallel to biological tissue, as will be explained in more detail later in connection with Table 5.1.

1.13 Complex Numbers in Electromagnetics (the Phasor Transform)

When the sources in EM systems are sinusoidal functions, a powerful method called the *phasor transform* is usually used to solve the EM field equations. In this method, the electric and magnetic fields and other functions of interest are transformed from functions of time t to functions of radian frequency ω. This is equivalent to saying that in the frequency domain, we know that the time shape of the wave is sinusoidal, so we only need to know the amplitude A and phase ϕ of the wave to completely understand what the wave is doing. Through this thought process, we have transformed our thinking from the *time domain* to the *frequency domain*. For example (starting with Equation 1.9), a wave in the time domain is given by

$$g(t) = A\cos(\omega t - \phi) \tag{1.19}$$

Since we know that the wave is sinusoidal [cos(ωt − ϕ)] and we know the frequency ω, the amplitude A and phase ϕ are all that are needed to fully understand this wave. The combinations of the transformed parameters (amplitude A and phase ϕ) in the frequency domain are called *phasors*. The phasor form of Equation 1.19 can be written as

$$\tilde{G} = Ae^{j\phi} = A\angle\phi. \tag{1.20}$$

The tilde (~) above the symbol indicates a phasor in the frequency domain. The phasor transform method consists of transforming all the EM functions and equations to the

frequency domain, solving the transformed
EM equations for the phasors, and then transforming the phasors back to the time domain
to obtain the desired EM field quantities.

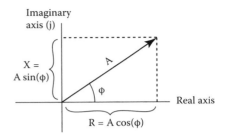

The phasor method is very advantageous
because the EM equations in the frequency
domain are algebraic equations, which are
easier to solve than the corresponding equations in the time domain, which are partial
differential equations. Often the math is most
easily done in a component form of the phasor.
Figure 1.34 shows the relationship between
the angular form of the phasor (Equation 1.20)
and the component form given below:

FIGURE 1.34
Geometric relationship between the angular and
component $(R + jX)$ form of phasor notation.

$$\tilde{G} = R + jX \qquad (1.21)$$

The real R and imaginary X parts of the phasor are related to the amplitude A and phase
ϕ components through simple trigonometry, as shown in Figure 1.34. The phasor is a complex number. A complex number is a number that has a part containing the square root of
–1, which we designate as $j = \sqrt{-1}$, and which is called an *imaginary number*. Some books
use the letter i instead of j. Note that $j^2 = -1$.

A complex number consists of a real part and an imaginary part. For example, 3 + j2 is a
complex number; 3 is the real part and j2 is the imaginary part. Electric field phasors, for
example, are often written as $E_r + jE_i$, where E_r is the real part and jE_i is the imaginary part.

The real and imaginary parts of phasors have important corresponding counterparts in
the time domain. The time-domain function that corresponds to the real part of the phasor
is 90° out of phase with the time-domain function that corresponds to the imaginary part of
the phasor. Thus, j in the frequency domain is associated with a 90° phase shift in the time
domain. The use of phasor transforms in EM field theory results in the definition of complex quantities such as complex permittivity and complex permeability (see next section).

The use of the phasor transform also results in the definition of *impedance* Z, which is the
ratio of phasor voltage to phasor current:

$$Z = \frac{\tilde{V}}{\tilde{I}} \cdot \text{(ohms)} \qquad (1.22)$$

The magnitude of the impedance gives the ratio of the magnitudes of the voltage and
current. The phase of the impedance tells by how many degrees the voltage leads the
current. Impedance is like resistance, in the sense that it opposes phasor current. That
is, for a given phasor voltage, the current is greater if the impedance is smaller. In fact, if
the impedance is a real number (no imaginary part, which means the phase is zero), the
impedance is a resistance. If the impedance has a phase that is not zero, it is complex and
is then different from resistance. Complex impedance takes into account the effects of
capacitance and inductance on phasor current. The important difference is the effect on
the loss of power. When current passes through a resistor (which has strictly real impedance), power is always lost. When current passes through a complex impedance, power is

not necessarily lost at all. In the case of inductances or capacitances, the power is simply delayed or stored, not lost. Sometimes we say the power is resonated, which means it passes between different parts of the circuit without being absorbed. Impedance is defined only in the frequency domain; it is not defined in the time domain.

1.14 Wave Properties in Lossy Materials

Waves propagating in free space (air) behave as described in Section 1.11; however, biological materials are lossy, and this loss changes the way the wave interacts with the material and, as a result, its propagation behavior. A material is lossy if $\sigma \neq 0$. Power will be deposited in the lossy material as a wave passes through it, thus causing loss to the propagating wave. If power is deposited in the material, the material will heat up.

To better understand how electromagnetic power is deposited in a lossy material, let us go back to Ampere's law. Equation 1.5 describes how a time-varying electric field produces a magnetic field. The phasor equivalent of Equation 1.5 (put in terms of $\mathbf{H} = \mathbf{B}/\mu$) is

$$\nabla \times \tilde{\mathbf{H}} = \tilde{\mathbf{J}}_c + j\omega\varepsilon_{bound}\tilde{\mathbf{E}} \tag{1.23}$$

The current term $\tilde{\mathbf{J}}_c$ is the conduction current, $\tilde{\mathbf{J}}_c = \sigma_c\tilde{\mathbf{E}}$, where the conductivity σ_c represents the mobility of free electrons in the material. The permittivity due to bound charges for lossy materials has both real and imaginary components:

$$\varepsilon_{bound} = \varepsilon_{b,real} - j\varepsilon_{b,imag} \tag{1.24}$$

Remember that the permittivity characterizes the interaction between the electric field and the bound charges in the material. This interaction results in an oscillatory motion of the bound charges. The portion of the motion that is 90° out of phase with the electric field is characterized by the real part of the permittivity, $\varepsilon_{b,real}$. This is a lossless interaction. But as the bound charges oscillate, they also heat up due to friction-like forces within the molecule and from molecules nearby. This motion component is in phase with the electric field and is represented by the imaginary part of the permittivity, $\varepsilon_{b,imag}$.

Considering all of these effects, Equation 1.23 can be written in the form

$$\nabla \times \tilde{\mathbf{H}} = \sigma_c\tilde{\mathbf{E}} + j\omega(\varepsilon_{b,real} - j\varepsilon_{b,imag})\tilde{\mathbf{E}} \tag{1.25}$$

Pairing the conduction term with the imaginary part of the permittivity (since both are in phase with the electric field) gives:

$$\nabla \times \tilde{\mathbf{H}} = (\sigma_c + \omega\varepsilon_{b,imag})\tilde{\mathbf{E}} + j\omega\varepsilon_{b,real}\tilde{\mathbf{E}} \tag{1.26}$$

The first term on the right-hand side of this equation, $(\sigma_c + \omega\varepsilon_{b,imag})\tilde{\mathbf{E}}$, represents the current that produces a loss (heat) in the material through movement of free charges and bound charges. Combining the two components in this lossy current term leads naturally to a definition of *effective conductivity*:

$$\sigma_{eff} = \sigma_c + \omega \varepsilon_{b,imag} \qquad (1.27)$$

Effective conductivity is the value normally given in tables of conductivity (such as in Appendix A). Many books simply use σ to represent effective conductivity and do not distinguish between conductivity (loss) from free and bound charges. Equation 1.27 also leads to the common definition of *effective conduction current density* (typically just called *conduction current*):

$$\tilde{J}_{c,eff} = \sigma_{eff}\tilde{E} \qquad (1.28)$$

The last term in Equation 1.26, $j\omega \varepsilon_{b,real}\tilde{E}$, is the displacement current, which represents the lossless portion of the oscillation of the bound charges. The higher the frequency ω, the faster the bound charges oscillate in the material and the larger the displacement current produced (as well as the loss due to the second term in Equation 1.27). The j indicates that the lossless displacement current (last term in Equation 1.26) is 90° out of phase with the lossy current term. Often $\varepsilon_{b,real}$ is simply designated ε in tables of dielectric properties.

Furthermore, in practice, the terms on the right-hand side of Equation 1.26 are often combined to write the equation as

$$\nabla \times \tilde{H} = j\omega \varepsilon_{complex}\tilde{E} \qquad (1.29)$$

where the *complex permittivity* $\varepsilon_{complex}$ is defined as

$$\varepsilon_{complex} = \varepsilon' - j\varepsilon'' = \varepsilon_0\left(\varepsilon'_r - j\varepsilon''_r\right) \qquad (1.30)$$

Here ε' represents the lossless component of the material properties and ε'' represents the combined components of the material interaction (due to both bound and free charges) that lead to loss. Comparing Equation 1.29 to Equation 1.26, the following relationships apply:

$$\varepsilon' = \varepsilon_{b,real} \qquad (1.31)$$

$$\omega\varepsilon'' = \sigma_c + \omega\varepsilon_{b,imag} = \sigma_{eff} \qquad (1.32)$$

In a similar fashion, *complex permeability* is defined as

$$\mu_{complex} = \mu' - j\mu'' = \mu_0\left(\mu'_r - j\mu''_r\right) \qquad (1.33)$$

where μ_0 is the permeability of free space and the quantity $(\mu'_r - j\mu''_r)$ is called the *complex relative permeability*. μ' describes the lossless interaction between the magnetic field and the magnetic dipoles of the material, while μ'' represents any loss associated with aligning the magnetic dipoles. For most applications, the human body is so weakly magnetic that we can assume $\mu_{complex} = \mu_0$, so $\mu'_r = 1$ and $\mu''_r = 0$.

The ratio of ε' and ε'' is very useful to understand the degree of loss in the material being considered. This ratio is called the *loss tangent* ($\tan\delta = \varepsilon''/\varepsilon' = \sigma_{eff}/\omega\varepsilon'$) or the *dissipation factor* because it represents the angle of $\varepsilon_{complex}$ in the complex plane. Often tables of electrical properties of materials give the loss tangent ($\tan\delta$) and relative permittivity ε'_r.

Remembering that $\varepsilon' = \varepsilon_0\varepsilon_r'$, effective conductivity can be found from these tables for specific frequencies of interest using $\sigma_{eff} = \omega\varepsilon_0\varepsilon_r' \tan\delta$.

When the loss tangent is large, this means the material is very lossy. This loss could be caused either by free charge conductivity, as in metals, or by a bound charge loss term in a lossy dielectric. Metals are very good conductors. In fact, in most bioelectromagnetic applications, they can be considered to be *perfect conductors*, which is approximated by letting $\sigma_{eff} \to \infty$.

Now the electric field **E** inside a *perfect* conductor ($\sigma_{eff} \to \infty$) must be zero, since otherwise the conduction current density $\tilde{J}_{c,eff} = \sigma_{eff}\tilde{E}$ would be infinite, an impossibility. Since the loss in a metal is proportional to the square of the electric field strength, the loss in a perfect conductor is theoretically zero. One of the few exceptions in bioelectromagnetics where metal cannot be considered to be a perfect conductor is the large MRI electromagnet coil. In that case, a very large (1–4 tesla) magnetic field must be produced, which requires very large currents in the coil. The small amount of resistance (lack of perfect conductivity) in the metal coil is enough to heat the coil and cause the loss of a great deal of power. Supercooling the coil with liquid nitrogen increases its conductivity (decreases its resistance) and allows it to function properly. An important interesting note on conductive materials is that the two parts of the conductivity (σ_c from movement of free charges and $\omega\varepsilon_{b,imag}$ from motion of bound charges) behave very differently as a function of frequency. The free charge component of conductivity σ_c is sometimes called the static conductivity because it is the only component present at DC (where $\omega = 0$), and because it does not change with frequency. Since it represents the mobility of free electrons, it does depend rather strongly on temperature, and hence the effectiveness of supercooling the MRI electromagnet. The bound charge component $\omega\varepsilon_{b,imag}$ depends strongly on frequency and (typically) to a lesser extent on temperature. At very high frequencies this term dominates, and at very low frequencies it is nonexistent. Thus, in theory, it would be possible to measure the individual components of the conductivity separately by measuring the effective conductivity at different frequencies. This is not generally done in measurements (which is why tables virtually always give effective conductivity at certain frequencies), but it is often done in simulation models* of electrical properties of materials that span a broad band of frequencies, particularly for mixtures.

When the loss tangent is small, the material has very low loss. These types of materials are good *insulators* or near-perfect *dielectrics*. Glasses and most plastics are low-loss materials. As described in Appendix A, plastics have been used by several researchers to create phantom models of the human head or torso with accurate electrical properties of the skin or bone. Skin, bone, and fat are low-water-content tissues and have relatively low loss. Adding only small amounts of conductive material (such as salts) to the plastic is sufficient to create phantom models of low-loss tissues. It is more difficult to use plastics to create phantoms for the lossier high-water-content tissues such as muscle and brain. Most researchers have used water-based materials to represent these tissues, although some models with plastics doped with carbon and other highly conductive materials have been used.

Many bioelectromagnetic materials are neither high nor low loss. Their loss tangents are moderate, and they fall in the great middle ground where both conduction and displacement currents play major roles. For these materials, it is very useful to remember that the conduction current term (using effective conductivity) accounts for the loss in the material and any subsequent heating effects, and that the displacement current term accounts for the out-of-phase motion of the bound charges, and is a lossless current.

---

* Such models include the Cole-Cole equation and the Debye equation.

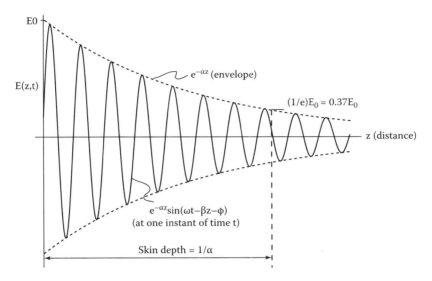

FIGURE 1.35
Propagating wave of the form $E(z, t) = E_0 e^{-\alpha z}\sin(\omega t - \beta z - \phi)$, shown at one instant of time. The peak magnitude of this wave in a lossy material decreases exponentially as a function of distance traveled. Its envelope is given by $e^{-\alpha z}$, and the effective penetration depth, or skin depth, is given by $1/\alpha$.

In a lossy material, the effective conduction current creates a loss in the material as the wave propagates, and the magnitude of the wave decreases exponentially. This is expressed by the formula $E(z,t) = E_0 e^{-\alpha z}\sin(\omega t - \beta z - \phi)$ for the electric field and is illustrated in Figure 1.35. The attenuation constant α represents how fast the wave attenuates as it propagates; it is given in units of nepers per meter. Actually, the neper (Np) is really dimensionless, so the units of α can also be thought of as loss per meter. The field inside the body as a function of propagation distance can be found from $E_0 e^{-\alpha z}$, where E_0 is the incident field just inside the surface of the body, and z is the propagation distance (how far the field has penetrated into the body).

Attenuation is a function of the permittivity and conductivity of the material, and of the frequency.* The attenuation constant α is calculated as

$$\alpha = \omega \sqrt{\frac{\mu'\varepsilon'}{2}\left(\sqrt{1 + \left(\frac{\sigma_{eff}}{\omega\varepsilon'}\right)^2} - 1\right)} \cdot (\text{Np/m}) \tag{1.34}$$

Almost all biomaterials are nonmagnetic, so $\mu' = \mu_0$. Note that increasing the effective conductivity or the frequency increases the loss, and hence the attenuation. The attenuation is used to determine an important parameter called *skin depth* (see Section 3.4.1). Skin depth is given by $1/\alpha$, and its units are in meters. Skin depth is the distance the wave propagates before its magnitude has dropped to $1/e = 0.37$, or about $1/3$ of its original value, where $e = 2.718$ is the base of the natural logarithm.

---

* The following two equations assume that there is no attenuation in the material due to magnetic dipole losses, so the permeability is entirely real ($\mu'' = 0$); this is the case for all human tissues and most other dielectrics.

The propagation constant also depends on permittivity, conductivity, and the frequency:

$$\beta = \omega \sqrt{\frac{\mu'\varepsilon'}{2}\left(\sqrt{1+\left(\frac{\sigma_{\text{eff}}}{\omega\varepsilon'}\right)^{2}}+1\right)} \cdot (\text{rad}/\text{m}) \tag{1.35}$$

The propagation constant indicates how much phase shift (in radians) occurs in the wave as it propagates through distance. Using this value for propagation constant β, the wavelength for lossy material can then be found from Equation 1.16.

Now let us try an example. For a frequency f = 433 MHz, the angular frequency, ω, is 2.72×10^9 rad/s. From Appendix A, Table A.1, we can read the values of relative permittivity for muscle $\varepsilon_r = 64.21$ and conductivity $\sigma = 0.9695$ S/m. Since we generally use 2/3 muscle to represent the human body, we multiply both ε_r and σ by 2/3, giving $\varepsilon_r = 42.81$ and $\sigma = 0.6463$ S/m. Then we can use Equation 1.34 to calculate $\alpha = 17.85$ Np/m and Equation 1.35 to calculate $\beta = 62.07$ rad/m. Then from Equation 1.16, which applies in both lossy and lossless cases, the wavelength $\lambda = 10.1$ cm. We can also calculate the velocity of propagation in 2/3 muscle at 433 MHz from Equation 1.14 (solving for v_p), which gives $v_p = 4.38 \times 10^7$ m/s. If the electric field just inside the body is $E_0 = 1$ V/m, the field at z = 10 cm inside the 2/3 muscle material will be $E_0 e^{-\alpha z} = 0.0168$. The power will be reduced by $e^{-2\alpha z}$ to 2.8% of its original value.

1.15 Boundary Conditions for Lossy Materials

Boundary conditions in lossy materials, similar to Equations 1.17 and 1.18, must be satisfied by the normal and tangential components of the phasor electric fields in the frequency domain. At the boundary, these are

$$\varepsilon_{\text{complex 1}} \tilde{E}_{n1} = \varepsilon_{\text{complex 2}} \tilde{E}_{n2} \tag{1.36}$$

$$\tilde{E}_{t1} = \tilde{E}_{t2} \tag{1.37}$$

Using the definition of complex permittivity, Equation 1.30, and the relationship between ε'' and σ_{eff} from Equation 1.32, the boundary condition on the normal E field, Equation 1.36, can be rewritten as

$$\left(\sigma_{\text{eff 1}} + j\omega\varepsilon_1'\right)\tilde{E}_{n1} = \left(\sigma_{\text{eff 2}} + j\omega\varepsilon_2'\right)\tilde{E}_{n2} \tag{1.38}$$

The first term on each side of this equation represents the effective conduction current density normal to the boundary on each side, and the second term represents the (90° out-of-phase) normal displacement current on each side. Thus, Equation 1.38 can be interpreted as

$$\tilde{J}_{c,\text{eff } n1} + \tilde{J}_{\text{displacement } n1} = \tilde{J}_{c,\text{eff } n2} + \tilde{J}_{\text{displacement } n2} \tag{1.39}$$

In equations containing complex numbers, the real part of the left-hand side must be equal to the real part of the right-hand side, and the imaginary part on the left-hand side must be equal to the imaginary part on the right-hand side. In light of Equation 1.39, Equation 1.38 shows that the normal conduction currents must be equal across the boundary, and the normal displacement currents must be equal across the boundary. This requirement holds because they are 90° out of phase with each other, and at all times the currents must be equal at the boundary.

1.16 Energy Absorption

In many electromagnetic field interactions, energy transfer is of prime consideration. For example, in hyperthermia for cancer therapy the electric field is transferred in the body into heat, which is the desired outcome of the therapy. For cell phones, the energy transfer must be below predefined regulations. The **E** field can transfer energy to electric charges through the forces it exerts on them, but the **B** field does not transmit energy to charges. The forces that **B** exerts on the charges can change their directions, but not their energy, because these **B**-field-exerted forces are always in a direction perpendicular to the velocities of the charges. The **B** field can, however, transfer energy through forces on permanent magnetic dipoles. Because biological tissue is mostly nonmagnetic (contains very few permanent magnetic dipoles), this latter effect is not prominent in EM biological interactions.

For sinusoidal steady-state EM fields, the power (time rate of energy) transferred to charges in an infinitesimal volume element Δv of a material is given by

$$P = \sigma_{eff} E_{rms}^2 \Delta v \ (W) \tag{1.40}$$

where ω is the radian frequency in radians per second (rad/s) (Section 1.8) and E_{rms} is the root mean square (rms) value (Section 1.13) of the electric field **E** *at that point*. Thus, Equation 1.40 is a *point relation*, because it applies only at the given point where **E** has that particular value. The time rate of energy change is called *power*, which has units of watts (W). The unit of energy is the joule (J). One watt is equal to one joule per second. Since Δv has units of m^3, the quantity $\sigma_{eff} E_{rms}^2$ has units of W/m^3, which is the density of absorbed power.

The *specific absorption rate* (SAR) is defined as transferred power divided by the mass of the object. *Specific* refers to the normalization to mass, and *absorption rate* to the rate of energy absorbed by the object. For sinusoidal steady-state EM fields the time-average SAR is given by

$$SAR = \sigma_{eff} E_{rms}^2 / \rho \ (W/kg) \tag{1.41}$$

where ρ is the mass density of the object in kg/m^3, which is close to 1.0 kg/m^3 for most biological tissues except for lung, which is about 0.347 kg/m^3. Again, Equation 1.41 is a point relation, so it is often called the *local* SAR. The *space-average* SAR for a body or a part of the body is obtained by calculating the local SAR at each point in the body and averaging over the whole body or the part of the body being considered.

From Equation 1.41, we see that the SAR varies directly with σ_{eff}. Generally speaking, tissue with higher water content, such as muscle, is more lossy for a given E field magnitude than drier tissue, such as bone and fat, as can be seen from Appendix A. Also, the higher the frequency, the higher σ_{eff}, due to the part of the loss caused by the motion of the bound charges. If the same field were present at both high and low frequencies, the power absorbed would be higher at the higher frequencies. The high frequencies attenuate more as they propagate through tissues than the low frequencies, however, so in general much less field is present at high frequencies. Detailed dosimetry (Chapter 5) is needed to determine how much field is present in order to determine localized SAR distributions.

THE BIOHEAT EQUATION

The temperature rise in tissue is determined by the rate of electromagnetic power deposition in the tissue (W_{em}) and the metabolic heating rate (W_m), as well as the thermal dissipation by conduction (W_c) and blood flow (W_b). The temperature rise is predicted by the bioheat equation:

$$\frac{\partial(\Delta T)}{\partial t} = \frac{1}{4186c}\left(W_{em} + W_m - W_c - W_b\right) \qquad (1.42)$$

where c is the specific heat of the tissue. Blood flow and conduction effects cause the temperature rise to be nonlinear. Typically the temperature rises linearly before vasodilatation increases the blood flow enough to start reducing the rate of the rise. In a 50 W hyperthermia treatment, this linear range lasts about 3 minutes. As more energy is absorbed in the tissue, the blood flow reaches a steady-state response and the temperature plateaus at a steady-state level.

Tumors typically have a high perfusion of blood at the periphery and low perfusion in the center. The steady-state blood flow in the normal tissue is also normally higher than in the tumor, thus cooling the normal tissue while the tumor continues to increase in temperature. This variant between the temperature of normal and tumor tissues typically begins at about 10 minutes into the treatment and reaches steady state at about 20 minutes. Thus, it is very important to maintain constant power deposition for an extended period of time in order to obtain therapeutic levels of temperature rise for hyperthermia treatment of cancer. The bioheat equation is also used to predict the rise of temperature for safety applications, implantable electronics (where the power deposited may be from conduction currents rather than electromagnetic absorption), and other applications.

1.17 Electromagnetic Behavior as a Function of Size and Wavelength

As illustrated in Figure 1.31, Maxwell's equations apply over an extremely broad frequency spectrum. The characteristic behaviors of EM fields, however, are significantly different for different frequency ranges. To be more specific, the characteristic behaviors depend on the size of the EM device or system as compared to the wavelength. Suppose that L is

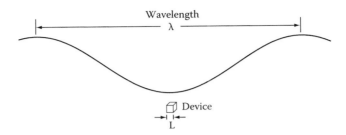

FIGURE 1.36
The wavelength is large compared to the size of the device.

the largest dimension of the device or system being considered and λ is the wavelength of the EM fields. Then the techniques used to solve Maxwell's equations and the characteristic behavior of the EM fields can be summarized in terms of three categories: $\lambda \gg L$ (Figure 1.36), $\lambda \approx L$ (Figure 1.37), and $\lambda \ll L$ (Figure 1.38), as shown in Table 1.1. For many typical devices, these three categories correspond to low frequency, medium frequency, and high frequency, respectively. However, the categories are really defined in terms of λ and L, regardless of the frequency. For example, $\lambda \gg L$ could apply at high frequencies when L is the largest dimension of a microcircuit. As a prelude to discussing the EM characteristic behaviors in each of these ranges, we need to point out that Maxwell's equations, Equations 1.3 and 1.5 (Faraday's and Ampere's laws), relate **E** and **B** in such a way that **E** and **B** are said to be *coupled*. That is, a time-varying **B** creates an **E** (Equation 1.3), and in turn the time-varying **E** creates **B** (Equation 1.5). Thus, when the fields vary with time, one cannot exist without the other, because each acts as a source of the other. When the fields do not vary with time, they are said to be *uncoupled*, because then they do not act as sources of each other.

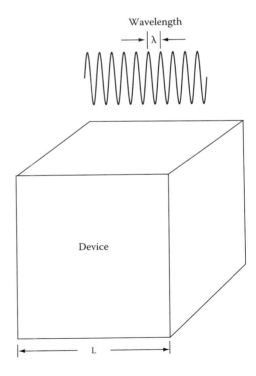

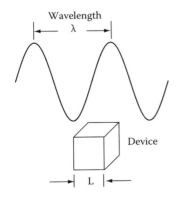

FIGURE 1.37
The wavelength and the size of the device are comparable.

FIGURE 1.38
The size of the device is large compared to the wavelength.

TABLE 1.1

Summary of EM Techniques and Characteristics as a Function of the Relationship Between Device or System Size L and Wavelength λ

When λ >> L (low frequency for most typical devices) (Figure 1.36)	Electric circuit theory and quasi-static EM field theory are used. Propagation effects are negligible. **E** and **B** are uncoupled. Energy is transmitted by wires and cables, but not in beams through the air.
When λ ≈ L (medium frequency for most typical devices) (Figure 1.37)	Microwave theory is used. Propagation effects dominate. **E** and **H** are strongly coupled. Energy is transmitted through cables, hollow waveguides, and beamed through the air.
When λ << L (high frequency for most typical devices) (Figure 1.38)	Optics and ray theory are used. Propagation effects dominate. Energy is beamed through the air and is not transmitted through metallic cables or along metallic wires, but can be transmitted through optical fibers.

When λ >> L (Figure 1.36), quasi-static EM field theory applies, which means that the spatial distribution of the EM fields over the extent of the device is the same as that of static fields, but the fields vary with time. For most human body applications, fields less than 1 MHz, or even 10 MHz, can be assumed to be quasi-static. **E** and **B** are said to be uncoupled when the ∂**E**/∂t and ∂**B**/∂t terms in Equations 1.3 and 1.5 are small enough to be neglected, which is often the case in this range. Therefore, a **B** field can exist without a corresponding coupled **E** field. Also, an **E** field can be produced by charges, as described by Equation 1.4, without a corresponding coupled **B** field. In some configurations in this range, though, a time-varying **B** field can still produce a significant **E** field. In this case, the **E** field is, of course, produced by the **B** field. An example of this is shown in Figures 1.12 to 1.14.

In this low-frequency range, a unique potential difference can be defined (see Section 1.2), and electric circuit theory (Kirchhoff's laws) is a good approximation to Maxwell's equations. Circuit theory is widely used in this range because it is much simpler than Maxwell's equations. Propagation effects are negligible in this range, and energy cannot be efficiently beamed through the air; it is transmitted along wires and through cables.

When λ ≈ L (Figure 1.37), that is, when the wavelength is of the same order of magnitude as the size of the system, EM field theory or microwave theory must be used. A unique potential difference cannot be defined in this range, except in special cases (see Section 3.5.1). Propagation effects dominate in this range, and **E** and **H** fields are described primarily in terms of propagating waves. **E** and **H** are strongly coupled. Neither **E** nor **H** fields can exist alone; the presence of one generates the other. Energy is typically transmitted along coaxial cables, through hollow pipes called waveguides, and beamed through the air. Calculations are often more difficult in this range than in the other two ranges, because Maxwell's equations must be solved without powerful approximations like circuit theory or ray theory.

When λ << L (Figure 1.38), EM fields are described by optical theory, except at extremely high frequencies, where theories appropriate to x-rays are used. Ray theory is an approximation that is often used in optical theory. Again, propagation effects dominate here, and **E** and **H** are strongly coupled together. One cannot exist without the other. A unique potential difference can be defined in special cases, but the concept is rarely used in this range. Energy cannot be transmitted along wires in this range, because they are too lossy; it is typically beamed through the air or transmitted through dielectric waveguides such as optical fibers. This range includes the infrared, visible light, ultraviolet light, and x-ray portions of the EM spectrum. The upper-frequency part of this frequency range is called the *ionizing radiation* range because the energy of the discrete energy-carrying packets of the EM wave, called photons, is great enough to cause ionization of atoms that the wave encounters, with corresponding danger to biological tissues. Below the ionizing frequency band, electromagnetic waves are said to be *nonionizing* because they do not break molecular bonds.

Table 1.1 and Figure 1.31 indicate the particular characteristics that apply to each general part of the spectrum. The ranges are only approximate representations, and the transition from one range to another is gradual, not abrupt. (Also, for some microcircuits, circuit theory may apply at higher frequencies than indicated in Figure 1.31 because L is very small for these circuits.)

THE VARYING BEHAVIOR OF A MICROWAVE OVEN DOOR AT TWO FREQUENCIES

Figure 1.39 illustrates how the behavior of a common device, the screen in a microwave oven door, can be explained in terms of the characterizations of Table 1.1. When an EM signal is incident on a metal plate containing an array of holes of diameter d, the signal is mostly blocked when d << λ; the signal interacts strongly and passes through when d ≈ λ; and the signal passes through the holes almost freely when d >> λ. The screen in a microwave oven door typically contains holes with diameters on the order of 2 mm. The wavelength of the microwave EM fields generated by the oven is typically about 122 mm (frequency is 2,450 MHz). The wavelength of the light produced by the lightbulb in the oven is on the order of 0.5 micrometers (μm). The screen thus blocks the microwave signal because for it, d << λ. On the other hand, the light passes through the holes almost freely because for it, d >> λ. The screen therefore nicely contains the microwave energy but lets the visible light pass through so that the contents of the oven can be observed.

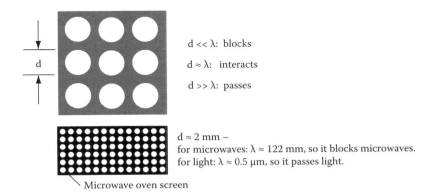

d << λ: blocks

d ≈ λ: interacts

d >> λ: passes

d ≈ 2 mm –
for microwaves: λ ≈ 122 mm, so it blocks microwaves.
for light: λ ≈ 0.5 μm, so it passes light.

Microwave oven screen

FIGURE 1.39
Behavior of the screen in a microwave oven door explained in terms of the characterizations of Table 1.1. The wave (propagating into or out of the paper) is incident on a plate filled with holes. If the holes are large compared to the wavelength (as in the case of light), the holes are transparent (and we can see into the microwave oven). If the holes are small compared to the wavelength (as in the case of microwaves), the plate acts like it is solid (and the waves stay inside the microwave cavity).

Because the characteristics of EM fields are so strikingly different in each of the three ranges described above, valuable insight can be gained by categorizing EM field interactions in terms of these ranges. Accordingly, we describe EM field behavior in each of these ranges in separate chapters, beginning in Chapter 2.

1.18 Electromagnetic Dosimetry

Electromagnetic dosimetry is the science of determining how much electric field (dose) exists for specific sources and environments. Dosimetry consists of two main parts. First, the incident **E** and **B** fields must be determined. Typically, these fields are determined either from the nature of the sources producing them or by measurements. Second, the **E** and **B** fields inside the object (for example, humans or other animals) must be determined, either by calculation or by measurement. The relationship between the incident EM fields and the internal EM fields is a strong function of the frequency of the incident fields, the size and shape of the body, and the electromagnetic properties of the body. Typically, different techniques are used to calculate and measure internal fields in each of the ranges described in Table 1.1. These techniques and typical results are described in subsequent chapters. In all cases, the relationship between the incident fields and the internal fields is very complicated.

In general, the penetration of incident fields into biological bodies decreases as frequency increases. This effect is illustrated by the graph in Figure 1.40, which shows the skin depth

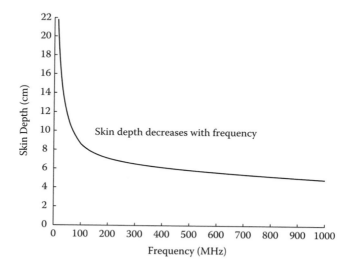

FIGURE 1.40

Skin depth as a function of frequency for a planewave incident on a dielectric halfspace having a permittivity and conductivity equal to two-thirds that of muscle tissue. The skin depth is the distance at which the wave has decreased to 1/e (about 1/3) of its original value due to attenuation (loss) in the tissue. Tissues with higher water content (e.g., muscle, brain) are more conductive and therefore have smaller skin depths (fields do not penetrate as far into these tissues). Tissues with lower water content (e.g., fat, bone) are less conductive and have larger skin depths. Lowering the frequency always increases the skin depth (and hence how far the wave will propagate in the body).

as a function of frequency for a planewave incident on a dielectric halfspace having a permittivity and conductivity equal to two-thirds that of muscle tissue. By dielectric halfspace, we mean that half of all space is filled with one dielectric, and the other half is filled with another dielectric (often free space), with a planar interface between the two. Two-thirds the permittivity and conductivity of muscle were used, because that is approximately the average of all the tissues in a typical human being. The skin depth is defined as the depth at which the EM fields have decreased to $1/e$ (i.e., 0.37) of their value at the surface of the body (see Sections 1.14 and 3.4.1). While the graph is for a dielectric halfspace, a similar effect occurs in humans and other animals. As the frequency is increased, the penetration generally becomes less and less. At optical frequencies, the penetration is very slight, and whatever effects the EM fields have on the body are primarily surface effects. Even at microwave frequencies, the penetration is relatively shallow (a few centimeters).

HOW ELECTROMAGNETIC FIELDS AFFECT THE BODY

Bioelectromagnetics is the study of how electromagnetic fields affect the body. This includes therapeutic applications as well as natural effects and safety concerns. Clearly, some biological effects are observed and utilized, which means there must be biological mechanisms with which fields affect the body. Interestingly, even after decades of extensive research, a few mechanisms are well quantified and understood, but others remain elusive and debatable. Seven basic effects are well understood and are described below.

Visual Phosphenes (Flashing Lights) (0.05 V/m Electric Field)

One of the most sensitive reactions to electromagnetic fields is the stimulation of apparent flashing lights, or *phosphenes*, in the eye. These are caused by a change of the synaptic potentials in the receptors or neurons of the retina. Originally it was assumed that electromagnetic fields caused direct stimulation of the visual cortex, but this theory has since been replaced by the synaptic potential model. The appearance of random flashing lights comes from retinal cells, which are stimulated to produce postsynaptic action potentials. The electromagnetic fields can change the presynaptic resting potential of the cell, and this potential can be greatly magnified in the cell's postsynaptic membrane. The most sensitive frequency for visual phosphenes is 20 MHz. At this frequency a 50 mV/m electric field on the retina is capable of producing visual effects. This is at least two orders of magnitude lower than what is required for other nerve stimulation thresholds.

Peripheral Nerve Stimulation (Sensation) (6 V/m Electric Field)

Nerve excitation can be caused by electromagnetic fields initiating depolarization of the neural membrane. This activates voltage-gated ion channels, which produce a propagating action potential down the length of the neuron, and thus a sensation. Nerve excitation occurs when the membrane is depolarized by about 15 to 20 mV and depends on the strength of the applied field, its polarization relative to the long axis of the neuron, and the length of time the excitation is in place. The most sensitive thresholds observed are for the largest myelinated nerves (about 20 μm in diameter), with a 6 V/m monophasic electric field pulse that is active for 100 to 200 μs and repeats every 2 ms. *(continued on next page)*

Direct Muscle Cell Excitation (6 to 12 V/m Electric Field)

Muscle cells are most easily stimulated by stimulating the motor neurons via the mechanism discussed above. They can also be directly stimulated by fields that are about ten times greater than those for nerve stimulation. A 6 V/m monophasic electric field pulse must be present for about 1 to 10 ms for skeletal muscle. A 12 V/m field is needed for stimulation of cardiac muscle.

Electroporation

Electroporation, where electric fields produce pores in a cell membrane (discussed in Section 1.1), occurs when a very strong magnetic (and associated electric) field is placed across a cellular membrane. A 50 V/cm pulsed electric field with a duration greater than 0.1 ms has been shown to produce reversible electroporation (pores close up when the field is removed), and a field of about 200 V/cm produces irreversible electroporation (pores do not close up with the field is removed). When the membrane potential is raised to about 800 to 1,000 mV, the individual pores can become enlarged and allow fluid (typically drugs) to flow. The induced electric field that is required to produce these membrane potentials depends on the specific configuration—if the field is induced external to the body, across an individual cell, across a group of cells, and so forth. Electroporation has been observed at field strengths as low as 50 to 300 V/cm.

Thermal Effects (Heating)

Probably the best understood of the electromagnetic interactions are thermal interactions with the body. These can be therapeutic (as in the case of heat therapy for muscular pains and hyperthermia for treatment of cancer). Tissue heating is caused by deposition of power from the ionic conductivity of the tissue as well as losses associated with motion of the molecules caused by the time variation of the electromagnetic field. Power deposition is measured by specific absorption rate (SAR), described in Section 1.14. An SAR value of 1 W/kg will require 1.1 hours of continuous exposure to achieve a 1°C increase in temperature. Typically this requires an internal electric field magnitude of 45 V/m at a frequency of 1 kHz, 35 V/m at 100 MHz, and 27 V/m at 1 GHz. Small gradients of temperature within the body can also produce sensation, even pain, such as those associated with hyperthermia at the bone-muscle interface (see Section 6.3).

Audio Effects (300 V/m Electric Field)

When a person's head is exposed to pulsed or very high-frequency electromagnetic fields that produce an SAR of 100 W/kg or higher, they may experience auditory effects. At 2.45 GHz, the only frequency where auditory effects have been studied extensively, this requires an electric field greater than 300 V/m in the head. This produces a small, transient thermal response that follows the pulse, which in turn produces a thermoelastic wave in the skull that can be perceived by the normal mechanical auditory mechanism of the ear. *(continued on next page)*

Magnetohydrodynamic Effects (Taste and Vertigo)

Very high magnetic fields (above 1.5 T) can induce a force on flowing ions (typically blood) in the body. This effect is most prevalent when the ions are flowing perpendicular to the magnetic field. The induced forces can cause strange taste sensations and vertigo. This effect also occurs when you move your head quickly in a strong magnetic field such as present in an MRI instrument. In addition, switched magnetic fields can cause drag on a conducting fluid (again, such as blood). This effect is quite small but measurable. For a 5T/s time-varying magnetic field, the pressure within the human vasculature system will change by less than 1%. Numerous other potential mechanisms for bioelectromagnetic effects have been proposed but are not yet fully understood. These are described in Chapter 6.

2

EM Behavior When the Wavelength Is Large Compared to the Object Size

2.1 Introduction

In Chapter 1 we explained that the characteristics of EM fields and their interactions with objects vary dramatically with the ratio of the wavelength of the EM fields to the object size. The purpose of this chapter is to describe in detail the behavior of EM fields and their interaction with objects when the wavelength is large compared to the size of the object. As indicated by Figure 1.31, the wavelength in free space is 300 m when the frequency is 1 MHz.

Consequently, for objects about the size of people, the wavelength will be large in comparison to the object size at frequencies below 1 MHz. (Note from Equation 1.15 that the wavelength varies inversely as the frequency.) Thus, the discussion in this chapter pertains mostly to the low-frequency region of the spectrum, including the commonly used power line frequencies of 50 and 60 Hz, and what is referred to as the extremely-low-frequency (ELF) band. The ELF band is designated as the band of frequencies from 30 to 300 Hz.

Many naturally occurring electric and magnetic effects are seen in this low-frequency range. Lightning and other static discharges are examples of capacitive discharge that are able to source very large amounts of current over a very small area and time, thus potentially doing significant damage or inflicting pain in that small area. This same effect can be used in a smaller-scale, more controlled fashion in electrophysiology, which uses pulsed fields to heal bone or soft tissue, stimulate damaged nerves or muscles, and reduce pain.

The nervous system is made up of a massive network of neurotransmitters and receptors. By interfacing with this natural electrical system using electrodes, one can receive neural signals for analysis or stimulate nerves to produce biomechanical function. Electromyography, electrocardiograms, cardiac defibrillation or pacing, and direct nerve stimulation are a few of the applications commonly seen in this frequency band. Most applications in the low-frequency band are for either stimulation or reception of nerves and tissues, rather than imaging. This is because low-frequency fields cannot be easily focused to provide good images. An exception to this is impedance imaging. The low- and high-water-content tissues of the body have very different electrical properties (i.e., insulators as opposed to conductors), and this is used to provide local images of the body. Some simple commercial devices such as electrical scales that measure your weight and estimate your body mass index use this concept. Finally, since the power grid relies on low-frequency (50 or 60 Hz) fields, much research has been done to determine if these fields, which are now so pervasive in our environment, are safe or hazardous in small doses. These applications are covered in this chapter to provide an understanding of how the theory of low-frequency fields is translated into practice.

2.2 Low-Frequency Approximations

A number of useful approximations can be made when the wavelength is large compared to the size of the object. These approximations are often called low-frequency approximations because, as explained earlier, the frequency is typically low when the wavelength is large compared to the object size.

An important low-frequency approximation is electric-circuit theory, which is an approximation to Maxwell's equations (Sections 1.4 and 1.5). Voltage and current are the principal variables in electric-circuit theory, which consists of Kirchhoff's voltage and current laws, along with some auxiliary relations. Fortunately, electric-circuit theory is much simpler than EM field theory. It typically involves two scalar (nonvector) functions, voltage and current, which vary with time. EM field theory, on the other hand, typically involves electric and magnetic fields, both of which are vector functions that vary with three space variables as well as with time. Life would be difficult indeed if we were required to solve Maxwell's equations for every situation for which circuit theory is commonly used.

The other main low-frequency approximation is called *quasi-static EM field theory*. In this approximation, the spatial variation of the **E** and **B** low-frequency fields is approximated as being the same as that of static (not varying with time) EM fields. This is a valuable approximation because the EM field equations are simpler for static fields than they are for time-varying fields. Quasi-static EM field theory is used whenever information about the **E** field or **B** fields is needed at low frequencies. For example, it is used to calculate **E** field patterns produced by high-voltage power lines. On the other hand, electric-circuit theory is used when the systems involve lumped elements like resistors, capacitors, inductors, and transistors, in which case the distribution of the **E** and **B** fields is usually not needed.

An example of a quasi-static **E** field is that produced by a sinusoidally time-varying low-frequency potential difference applied between two parallel metallic plates (a capacitor). This potential difference produces an **E** field between the plates that is similar to that shown in Figure 1.3, but the pattern as a whole varies sinusoidally with time. That is, the magnitude of the vector shown at each point changes sinusoidally with time in synchrony with all the other vectors; however, the relative spatial pattern remains the same. This sinusoidal variation includes negative values (refer to Section 1.9 and Figure 1.24), which means that the vectors reverse directions periodically. In the example of the parallel-plate configuration, the **E** fields are found from the quasi-static approximation by solving the static EM field equations and then letting the resulting **E** vary sinusoidally with time, which is much easier than directly solving the time-varying EM field equations.

An important consequence of quasi-static field theory is that the **E** and **B** fields can exist independently, or they are said to be *uncoupled*. This can be seen from Maxwell's equations. As indicated by Equations 1.3 and 1.4, **E** can be produced by either a time-changing **B** or a distribution of charges, or both. It is important to note that a time-changing **B** is not necessary to produce the **E**, just charges. In fact, in the static case, $\partial \mathbf{B}/\partial t$ is zero and will not contribute to the **E**. In the quasi-static case, $\partial \mathbf{B}/\partial t$ is often (though not always) small enough to be neglected in comparison to contributions to **E** due to the charges. Thus, a quasi-static **E** field can be produced by a charge distribution independently of any **B**, as indicated more

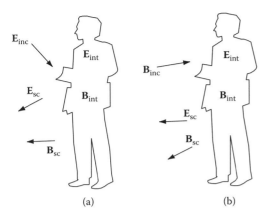

FIGURE 2.1
The internal and scattered **E** and **B** fields due to (a) an incident **E** alone and (b) an incident **B** alone. The internal and scattered fields are generally not the same in (a) and (b).

specifically by Equation 1.4. Similarly, from Equations 1.5 and 1.6, we see that when the $\partial E/\partial t$ can be neglected in Equation 1.5, a **B** can be produced by a current independently of any **E**. On the other hand, in the high-frequency case, when the $\partial B/\partial t$ and $\partial E/\partial t$ cannot be neglected, Equations 1.3 and 1.5 show that **E** and **B** are coupled together, and one cannot exist without the other.

As a consequence of the quasi-static approximation, when the wavelength is large compared to the object size, we can describe the interaction of EM fields with objects in terms of the two cases described in Figure 2.1. A typical situation consists of a set of sources that produces known EM fields in which an object is then placed. Some definitions are helpful: The *incident* **E** and **B** are the **E** and **B** fields produced by the sources without the object present. The *internal* **E** and **B** are the fields that exist inside the object. They consist of the original incident fields modified by the presence of the object. The *external* fields are the fields outside the object. They consist of the original incident **E** and **B** plus the *scattered* **E** and **B**, which are the fields produced by the presence of the object in the incident fields. Sometimes the internal fields are of primary interest, and sometimes the scattered fields are of primary interest. At low frequencies, the internal fields are usually of most interest.

Because the **E** and **B** are uncoupled at low frequencies, the internal fields can be found by finding the internal **E** and **B** due to the incident **E** alone, and then finding the internal **E** and **B** due to the incident **B** alone, as illustrated in Figure 2.1. The total internal **E** is the vector sum of the internal **E** in (a) and the internal **E** in (b). A similar procedure can be used to find the internal **B** and the scattered **E** and **B**. Typical examples of each kind of internal field are given in the next sections.

2.3 Fields Induced in Objects by Incident E Fields in Free Space

Figure 2.2 shows the calculated internal **E** in a section of a simple two-dimensional model of a dielectric prolate spheroid placed in the uniform **E** (similar to that shown in Figure 1.3) that existed between two metallic plates before the object was inserted. (A prolate spheroid has the shape of an egg.) When the spacing between the plates is very large compared to the length of the body, this configuration produces approximately the same field pattern that occurs when the body is placed in a uniform **E** field in free space. In this numerical calculation (a finite-difference frequency-domain solution of Maxwell's equations), we

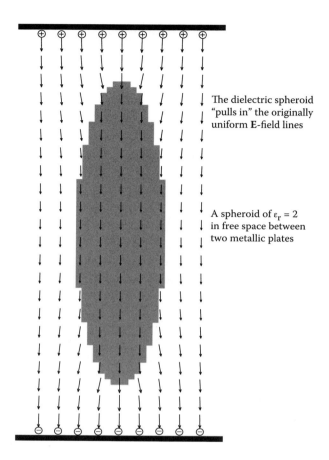

The dielectric spheroid "pulls in" the originally uniform E-field lines

A spheroid of $\varepsilon_r = 2$ in free space between two metallic plates

FIGURE 2.2
Calculated **E** fields in a two-dimensional model of a prolate spheroid between two metallic plates, approximating a spheroid placed in a uniform **E** field. The **E** fields were calculated on a finer grid but displayed on a coarser grid to show more clearly the overall field pattern. (The fields on the finer grid are shown in Figure 2.3.) The electric field tends to be pulled in toward the spheroid, where its normal component is reduced at the boundary. This type of model indicates roughly what would happen to a human with the electric field oriented from head to toe.

did not make the spacing between the plates larger because that requires more computer memory, but the characteristics of the field patterns are approximately the same as those for a prolate spheroid placed in a uniform **E** in free space.

In Figure 2.2, the relative permittivity (see Section 1.6) of the prolate spheroid is $\varepsilon_r = 2$, which is a comparatively small value, since the relative permittivity of free space is 1. We used this comparatively small value of relative permittivity for illustrative purposes, because larger values make the internal fields too small to be seen. The presence of the dielectric object perturbs the originally uniform **E** field in a way that can be thought of as the dielectric pulling in the **E** field vectors. In smooth prolate spheroids, the magnitude of the **E** field inside is uniform in space, but in this model, in which the outer boundary is stair-stepped (a consequence of the numerical calculation method), **E** is only approximately uniform because of the stair stepping. The **E** vectors in Figure 2.2 were calculated on a finer grid, but displayed on a coarser grid to show more clearly the overall field pattern.

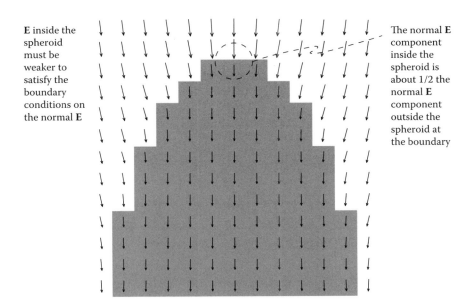

E inside the spheroid must be weaker to satisfy the boundary conditions on the normal **E**

The normal **E** component inside the spheroid is about 1/2 the normal **E** component outside the spheroid at the boundary

FIGURE 2.3
The field pattern around the top portion of the spheroid in Figure 2.2 as calculated on the finer grid. From this picture, it is clearer that the normal components of the field are reduced at the boundary, but the tangential components remain the same.

Figure 2.3 shows an enlarged view of the field pattern around the top portion of the spheroid as calculated on the finer grid. Note that the **E** field is weaker inside the object than outside. For quasi-static **E** fields, the **E** field is virtually always weaker in materials with higher permittivity. The electric flux density **D**, on the other hand, is controlled only by the source of the field (the distribution of charges on the metal plates in a capacitor, for instance). For the models shown here, **D** is uniform throughout. This means that the sources (charge distribution) are uniform, thus creating a uniform set of **D** flux lines. The effect that these flux lines have in a region on a charge (i.e., the force on the charge) is measured by **E**. The dielectric material introduces electric dipoles that rotate in a direction to partially cancel the incident **E** field, and hence reduce the force (and **E**) a charge would experience.

The fact that **E** is weaker inside the dielectric than in the surrounding air can be mathematically described in terms of the boundary conditions. As explained in connection with Equation 1.17 and Figure 1.32, the boundary conditions on E require that the normal component of **E** must be discontinuous at the boundary between two dielectrics by the ratio of the permittivities. In this case, if ε_1 is the permittivity of air and ε_2 is the permittivity of the spheroid, then Equation 1.17 requires that $E_{n2} = (\varepsilon_1/\varepsilon_2)E_{n1} = (1/2)E_{n1}$ *at the boundary.* From Figure 2.3, you can see that at the top of the spheroid, the **E** fields are approximately normal to the boundary, and that the normal component in the dielectric is about half that of the normal component in air. These fields are not the fields right at the boundary, but they are close enough to the boundary that they approximately satisfy the relationship required by the boundary conditions.

Figures 2.4 and 2.5 show the **E** field patterns for a prolate spheroid with its long axis perpendicular to the originally uniform **E** field in which it is placed, essentially turned 90° with respect to the orientation of Figure 2.2. The behavior is similar, with the **E** fields weaker in the dielectric than in the surrounding air above and below the object.

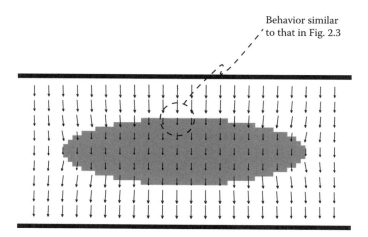

FIGURE 2.4
Calculated **E** fields in a model similar to that used in Figure 2.2, but with the long axis of the spheroid perpendicular to the originally uniform **E** in which it was placed. This type of model indicates roughly what would happen to a human with the electric field oriented from shoulder to shoulder. The fields are shown in more detail in Figure 2.5.

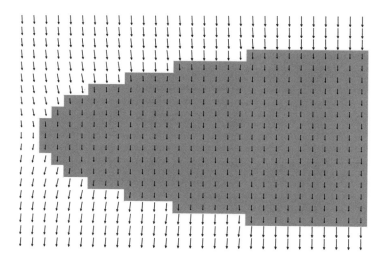

FIGURE 2.5
The field pattern near the left half of the spheroid in Figure 2.4 as calculated on the finer grid.

If the permittivity of the dielectric were increased to a much larger value, the boundary conditions would require the normal **E** in the dielectric at the boundary to be much smaller than the normal **E** in the air at the boundary. This illustrates the following important characteristic of **E** fields in this low-frequency region of the spectrum: the **E** fields inside dielectric objects with relatively high permittivities are usually much smaller than the **E** fields in the surrounding air. This behavior will be illustrated and explained more fully in subsequent sections.

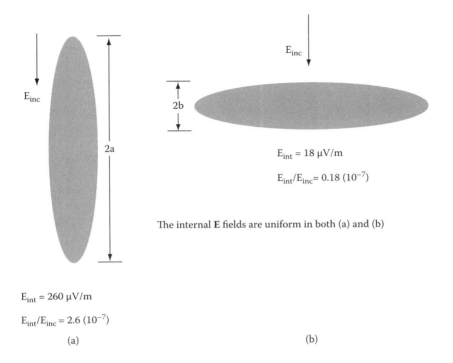

$E_{int} = 18 \ \mu V/m$

$E_{int}/E_{inc} = 0.18 \ (10^{-7})$

The internal **E** fields are uniform in both (a) and (b)

$E_{int} = 260 \ \mu V/m$

$E_{int}/E_{inc} = 2.6 \ (10^{-7})$

(a) (b)

FIGURE 2.6
Comparison of the internal fields in two spheroids, (a) with the incident **E** parallel to the long axis of the spheroid, and (b) with the incident **E** perpendicular to the long axis of the spheroid. In both cases the incident **E** is 1 kV/m, the conductivity is 0.067 S/m, the frequency is 60 Hz, a = 0.875 m, and b = 0.138 m. The internal **E** fields were calculated in three dimensions using a long-wavelength approximation to Maxwell's equations. Data for cross sections of similar two-dimensional models are shown in Figures 2.2 to 2.5.

Three-dimensional quasi-static solutions of Maxwell's equations in spheroidal coordinates for a prolate spheroidal model give more specific results about the internal fields in spheroidal objects, as illustrated in Figure 2.6, in this case for objects with more realistic relative permittivities near those of tissues. In both cases shown, the internal fields are six or seven orders of magnitude smaller than the incident field in which the objects were placed. This is attributable to the high permittivity and conductivity of the objects, which are approximate averages of all the tissues in the human body. Although the relative permittivity is of the order of 10^6, at this low frequency the conductivity probably dominates in determining the internal fields.

The ratio of the internal to incident **E** field for the case shown in (a) is almost fifteen times greater than that for the case shown in (b) even though the object is the same in each case. This difference can be explained in terms of the boundary conditions. The incident field in (b) is mostly normal to the dielectric-air interface over a much larger portion of the surface of the body than it is in (a). Because the boundary condition as stated in Equation 1.17 requires the internal normal field to be smaller than the external normal field by the ratio of the permittivities, the internal field is smaller in (b) because the boundary condition forces it to be that way over a larger portion of the surface at the boundary.

2.4 E Field Patterns for Electrode Configurations

In the last section, we discussed the **E** field effects of objects placed in uniform **E** fields in free space, as approximated by objects placed between metallic plates where the field was uniform before the object was placed there. In this section, we discuss **E** field effects produced by electrode configurations, such as those used in *in vitro* experiments.

2.4.1 Capacitor-Plate Electrodes

An important characteristic of **E** field interactions is illustrated in Figure 2.7, which shows the **E** field pattern produced by a current source (not shown) connected between two metallic plates (a capacitor) in the presence of a slender metallic object. The presence of the object perturbs the otherwise uniform **E** field significantly, particularly near the sharp corners at its end. Figure 2.8 shows an enlarged view of the area around the object. This is an example of a general characteristic, that sharp corners and objects cause a concentration and enhancement of **E** fields, as required by the boundary conditions. This can be explained as follows: the high conductivity of the metal makes the **E** fields inside the metal very small. Since the tangential components of **E** must be equal at the boundary, the tangential **E** in the air at the boundary must therefore be very small, approximately zero. This means that the **E** field in the air at the metal boundary must everywhere be normal to the metal. For the **E** to be normal to the metal everywhere around the corners means that the **E** field vectors must be crowded in, thus making the magnitude of **E** large. The more slender and pointed the object, the more the resulting concentration of **E** fields. Anyone who has inadvertently left a fork or other pointed metal object in a microwave oven has certainly seen the effect (sparks!) of this field concentration.

The fact that **E** fields concentrate at the corners of metal objects impacts a number of biological applications. As explained in Section 1.12, electric field probes must be made from very small metal components, with the remainder of the probe made from high-resistivity plastic. Also, electrodes used for delivering pulsed electromagnetic fields (PEMFs) are rounded on the corners to provide uniform pulse delivery.

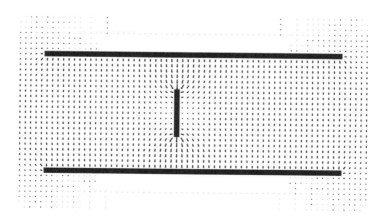

FIGURE 2.7
Calculated two-dimensional **E** field pattern for a slender metallic object placed in the (originally uniform) **E** between capacitor plates. The fields concentrate on the ends of the metallic object.

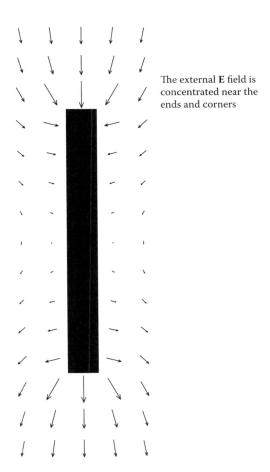

The external **E** field is concentrated near the ends and corners

FIGURE 2.8
Expanded view of the region around the slender metallic object in Figure 2.7. Consistent with the field concentration near the ends of the object, it is clear that the fields become normal (perpendicular) to the metal object. This is because tangential electric fields on the surface of metal are approximately zero.

A similar concentration of **E** fields is produced by dielectric objects, as illustrated in Figure 2.9, which shows a close-up view of the same capacitor arrangement as in Figure 2.7, but with a slender dielectric object having a relative permittivity of 100, instead of a metallic object. The dielectric has the effect of pulling the **E** field lines into it (as in the spheroidal example in Section 2.3), thus distorting the otherwise relatively uniform fields and concentrating the **E** fields around the corners.

An interesting example of this general effect would be you standing on the ground in a thunderstorm. The highly charged clouds produce a strong **E** field between the clouds and the earth, similar to the **E** produced between the metallic plates in Figure 2.7. You would be like the object between the plates, concentrating the **E** field at the ends of your body. If you pointed a finger to the sky, the resulting more slender and pointed object would increase the concentration of the **E** field, perhaps enough to cause ionization and breakdown of the air or, in other words, lightning. Lightning rods are sharply pointed for this reason. Golfers (whose upswung metal clubs electrically appear very much like lightning rods), farmers on tractors, and horseback riders are all particularly prone to lightning strikes because of their tall pointy outlines and the high probability that they are the tallest conducting

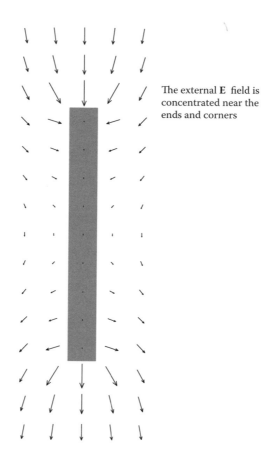

The external **E** field is concentrated near the ends and corners

FIGURE 2.9
Expanded view of the region around a slender dielectric object with a relative permittivity of 100 placed between the capacitor plates in Figure 2.7. The effect on the field is similar to that for a metal object shown in Figure 2.8. Fields concentrate at the ends and corners and are nearly perpendicular to the surface of the object.

objects in the vicinity. Ben Franklin's kite and key experiment would also accomplish the same purpose of concentrating the E field, particularly if the cotton kite string were wet and conductive. This is, of course, something that children today are taught is very unsafe! The best strategy in a thunderstorm is to make yourself as round and blunt as possible. Lying prone on the ground is not a good thing to do, because you could be injured or killed by currents in the ground produced by the charged clouds. Instead, it is typically recommended that you crouch and put one hand on the ground.

Another interesting effect occurs when layers of different materials are placed between capacitor plates. The flux density **D** is constant across all of the layers, but the effect measured by **E** is very different in the different layers. Figure 2.10 shows the **E** fields that occur in two layers, one with a permittivity and conductivity similar to those of muscle, and the other similar to those of fat. The **E** field in the fat is obviously much larger than the **E** field in the muscle. This occurs because the **E** fields in the two media are almost uniform and normal to the interface between the two materials. The boundary condition (Equation 1.38) requires that

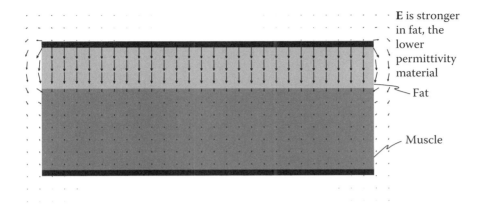

FIGURE 2.10
Calculated two-dimensional **E** field pattern at one instant of time for two layers of material having the electrical properties of fat and muscle placed between capacitor plates at 100 kHz.

$$(\sigma_f + j\omega\varepsilon_f)\tilde{E}_{nf} = (\sigma_m + j\omega\varepsilon_m)\tilde{E}_{nm} \qquad (2.1)$$

at the boundary, where the subscript f refers to the fat and the subscript m to the muscle.

For the conditions shown in Figure 2.10, $\omega = 2\pi \times 10^5$ radians/s, $\sigma_f = 0.026$ S/m, $\varepsilon_f = 180\varepsilon_0$, $\sigma_m = 0.477$ S/m, and $\varepsilon_m = 5,758\varepsilon_0$. Since the conductivity and permittivity of muscle are both much larger than those of fat, the boundary condition requires that $\tilde{E}_{nm}$ be much smaller than $\tilde{E}_{nf}$. As a consequence, the fat is heated much more than the muscle, as described by Equation 1.41. Although σ_f is smaller than σ_m, the E_{rms} in the fat is greater than that in the muscle, and since the specific absorption rate (SAR) varies as E_{rms}^2, the SAR in the fat is considerably greater than in the muscle. This effect is clearly seen when a fatty pork chop (or strip of bacon) is reheated in a microwave oven. The fat of the pork chop is sizzling long before the meaty portions are even warm.

This simple example illustrates an important characteristic behavior: **E** fields normal to the interface between a high-permittivity material and a low-permittivity material produce high **E** fields in the low-permittivity material. This characteristic behavior explains why some EM applicators used to heat tumors for cancer therapy cause the fat to overheat; they produce **E** fields normal to the fat-muscle interface.

2.4.2 Displacement Current

The capacitor of Figure 2.11 illustrates another important concept: *displacement current*. The **J** + ε ∂**E**/∂t terms in Equation 1.5 represent total current density. Equation 1.5 is called a *point relation* because it describes the relationship between the fields at each point in the system (not across the system as a whole). The conduction current term **J** is current density due to the movement of charges (see Section 1.6), both at points in free space and at points in materials. The ε ∂**E**/∂t term in free space does not involve movement of charge at all (since there are no free charges in free space), but it does have the same units as **J** (A/m²), and it has the characteristics of a current density. It is called *displacement current density*.

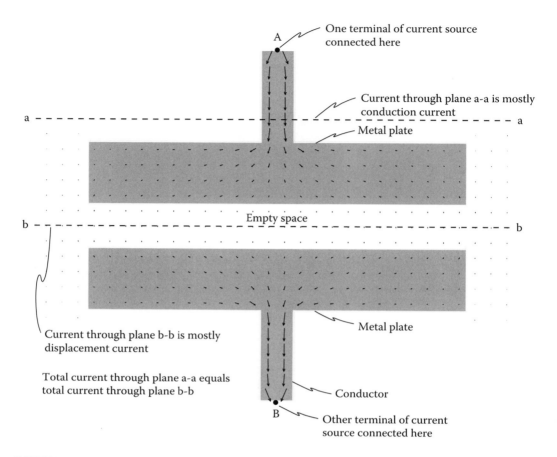

FIGURE 2.11
Calculated total current density (conduction plus displacement) in a two-dimensional model of a parallel-plate capacitor. The wires coming in are shown at the top and bottom, and the conductors are thick (shown in gray). Empty space is between and around the conductors. A sinusoidal current source is connected between points A and B. The current density is shown at an instant of time when the current source is maximum. The total current through each plane is equal, so the regions with a smaller cross section (the wires) have a larger current density, and regions with a larger cross section (plates and the area between them) have a lower current density.

Including this term was the triumph of James Clerk Maxwell in formulating the famous equations named after him (Sections 1.4 and 1.5). In materials, $\varepsilon\, \partial E/\partial t$ includes the effects of polarization (rotating or polarizing the electric dipoles as described in Section 1.6), but not those of free charge, which are included in the conduction current.

Displacement current is very important in capacitors (and also in how fields behave within the low-loss materials of the human body, such as fat, bone, and cartilage). Figure 2.11 shows a vector plot of the *total* current density $J + \varepsilon\, \partial E/\partial t$ at a snapshot in time when the sinusoidal current source is a maximum. The conduction current density spreads out in the thick conducting plates, as shown in more detail in Figure 2.12. Figure 2.13 shows the current density only in the air between and around the plates, plotted to a different scale. All of the current supplied by the current source must pass through all planes parallel to a-a in Figure 2.11. This means that the total current (current density times cross-sectional area) passing through *any* plane parallel to a-a in Figure 2.11 is the same and is equal to the current supplied by the current source. In plane a-a, the current consists of

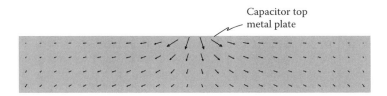

FIGURE 2.12
A close-up view of the current density in the top conductor of the capacitor in Figure 2.11 (plotted to a different scale). Here the current is predominantly conduction current.

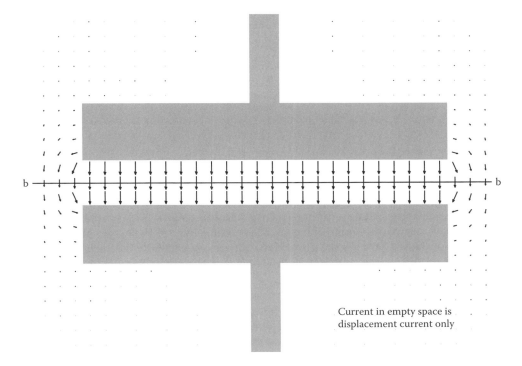

FIGURE 2.13
Calculated current density in the capacitor of Figure 2.11, but showing only the current density in the empty space surrounding the conductors, plotted to a different scale than Figure 2.11. Here the current is composed of displacement current only.

mostly the conduction current in the wire; the displacement current in the air is negligible. The total current passing through plane b-b in Figure 2.13, however, consists only of displacement current, since there are no free charges present there. This example shows how displacement current effectively continues current through space where no charges are present, making the total current (conduction plus displacement) continuous.

2.4.3 *In Vitro* Electrode Configurations

An experimental configuration that is sometimes used in the laboratory to expose solutions to **E** fields is illustrated in Figure 2.14, which shows the **E** fields produced by two wire electrodes placed in a saline solution in a container with nonconducting walls (such

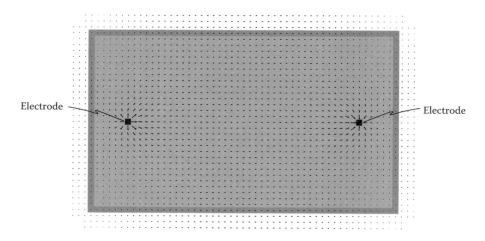

FIGURE 2.14

Calculated **E** fields in a two-dimensional model of two wire electrodes (the black squares) placed in a saline solution in a nonconducting container. A 60 Hz sinusoidal current source is connected between the two electrodes. The **E** fields are shown at an instant of time when the current source is a maximum.

as a petri dish). The figure shows the calculated **E** field in a two-dimensional model. These results are similar to those of a three-dimensional model looking down on the container from the top, with the fields shown in a plane perpendicular to the electrodes. Figure 2.15 shows the fields just in the saline, to give a closer view of the field pattern. The **E** fields are normal to the electrode surfaces, and are somewhat uniform in the central region between the electrodes, but are quite nonuniform in the regions around the electrodes. Figure 2.16 shows a close-up view of the region around the upper-left corner of Figure 2.14. **E** fields are produced both in the container walls and in the space outside the walls.

A side view of the electrode configuration of Figure 2.14 is shown in Figure 2.17, which shows the calculated **E** fields in another two-dimensional model. These two-dimensional results are similar to those that would be obtained from a three-dimensional model in a plane passing through the electrode centers. The field pattern in this plane is quite uniform between the electrodes, but again is variable near the ends of the electrodes.

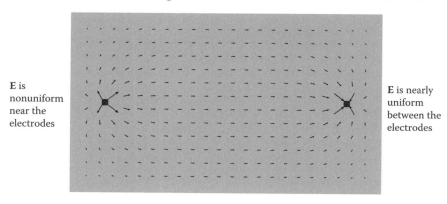

FIGURE 2.15

A close-up view of the **E** fields in just the saline of Figure 2.14. The **E** fields are interpolated and plotted on a coarser grid than in Figure 2.14 to show the field pattern more clearly. The fields are nearly uniform between the electrodes but experience significant fringing and nonuniformity near the electrodes.

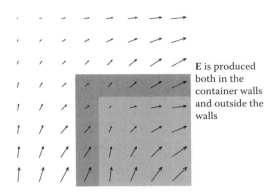

E is produced
both in the
container walls
and outside the
walls

FIGURE 2.16
A close-up view of the **E** fields in the region of the upper left corner of Figure 2.14.

E is variable near
electrode ends but
mostly uniform
between electrodes

FIGURE 2.17
Calculated **E** fields in a two-dimensional model of two wire electrodes (shown in black) placed in a nonmetallic container of saline (a side view). A 60 Hz sinusoidal current source (not shown) is connected between the two electrodes. The fields are nearly uniform between the electrodes but have fringing at the ends.

Figure 2.18 shows the effect of a nonconducting "bump" possibly representing a biological sample in the bottom of the nonconducting container. The nonconducting bump forces the current to flow around the bump in the saline, thus distorting the **E** field pattern in its vicinity, as shown more clearly in the close-up view of Figure 2.19. The **E** field inside the bump is higher than in the surrounding saline for the same reason the fields in the fat of Figure 2.10 are higher than those in the muscle.

Another electrode configuration is shown in Figure 2.20, where a plate electrode has replaced the left electrode of Figure 2.14. A closer view of the fields around the electrodes (Figure 2.21) shows how the **E** field is normal to the plate in the region just to the right of the plate, but is not much different from the pattern in Figure 2.14 around the wire electrode. Again, note the concentration of **E** fields at the tip of the wire electrode compared to the plate.

Two plate electrodes are shown in Figure 2.22, with a closer view shown in Figure 2.23. The **E** field pattern between the plates is much more uniform than in any of the other configurations shown previously. This would be expected from the geometry of the configurations because the **E** fields must be normal to the metallic electrodes. Plate electrodes would obviously be a better choice in any experiment in which a uniform **E** field is desired.

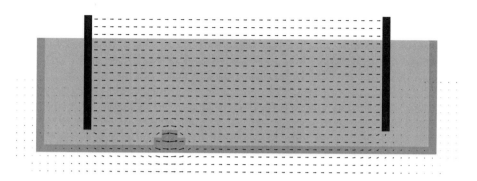

FIGURE 2.18
Calculated **E** fields in the model of Figure 2.17, but with a nonconducting bump on the bottom of the nonconducting container. The field is higher in the lower-conductivity bump. The bump perturbs not only the fields
inside it, but also the fields in the nearby saline.

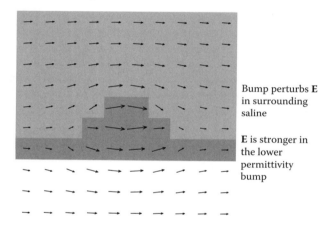

Bump perturbs **E**
in surrounding
saline

E is stronger in
the lower
permittivity
bump

FIGURE 2.19
A close-up view of the **E** fields around the bump of Figure 2.18.

This two-dimensional simulation does not show the effect of the corners of these plates
in a face-on view, but since we know that fields concentrate at the corners of plates, much
as they do at the tips of wires, the most uniform fields will be produced by plates with
rounded corners (or by round plates).

Figures 2.24 and 2.25 show what happens when a model of a nonconducting membrane
enclosing a saline interior (a rough model of a biological cell) is placed between the plate
electrodes of Figure 2.22. The nonconducting membrane forces current to flow around it,
and the **E** fields in the membrane are much higher than in the surrounding saline, for the
same reasons as explained in connection with Figure 2.10. The fields in the saline region
inside the membrane are essentially zero, as seen in Figure 2.25.

The field patterns shown in Figures 2.14 through 2.25 illustrate the nature of the **E** field
patterns produced by various electrode configurations in solution. In some places the **E**
fields are approximately uniform over a limited region, but near the electrodes they are
not close to being uniform, especially near small or pointed electrodes. These effects must
be carefully taken into account when doing experiments in which the dosimetry (strength
or distribution) of the **E** fields is important.

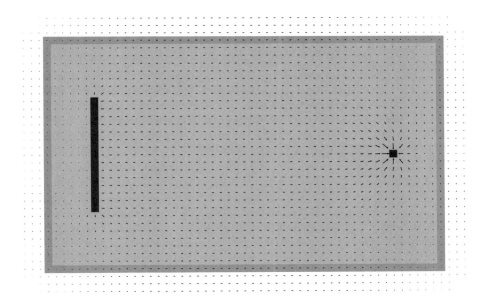

FIGURE 2.20
Calculated **E** fields in the model of Figure 2.14, but with the left wire electrode replaced by a plate electrode. The fields are still nearly uniform between the electrodes, but fringing fields are seen near the wire and the edges of the plate. The fields near the plate are more uniform than the fields near the electrode.

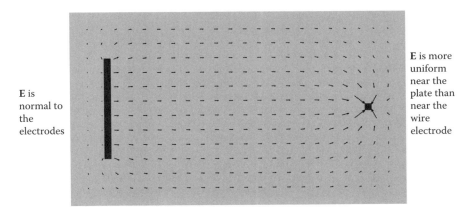

FIGURE 2.21
A close-up view of the **E** fields in the region near the electrodes of Figure 2.20. The **E** fields are interpolated and plotted on a coarser grid than in Figure 2.20 to show the field pattern more clearly.

2.5 Electrodes for Reception and Stimulation in the Body

Electrodes can be used two basic ways in medical applications. They can be used for *reception*, to receive signals from the body, such as for diagnosing carpal tunnel and other nerve disorders or for impedance imaging and monitoring. Electrodes can also be used

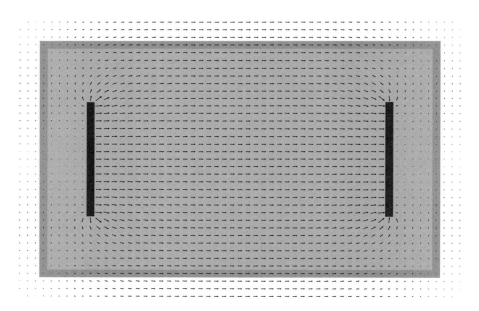

FIGURE 2.22
Calculated **E** fields in the model of Figure 2.20, but with the right wire electrode replaced by a plate electrode.

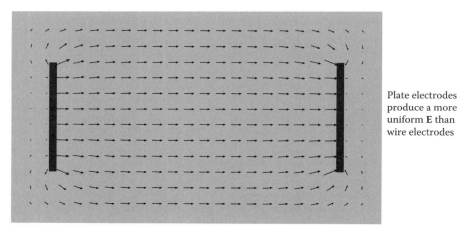

Plate electrodes
produce a more
uniform **E** than
wire electrodes

FIGURE 2.23
A close-up view of the **E** fields in just the saline of Figure 2.22. The **E** fields are interpolated and plotted on a coarser grid than in Figure 2.22 to show the field pattern more clearly. The uniformity of the fields is improved by using two plate electrodes, but fringing is still seen near their edges.

for *stimulation*, for example, delivering shocks for cardiac defibrillation or pacing, delivering pulsed electromagnetic fields for pain therapy and healing of muscle, bones, and nerves, or for stimulating the brain for control of Parkinson's disease or mental disorders. They can also be used for invasive heat treatments such as cardiac ablation, and for measuring the dielectric properties of body tissues and other materials. Electrodes for either stimulation or reception can be used over very large regions of the body, smaller regions, or very localized areas. They can be invasive, external, *in vitro*, or completely separated

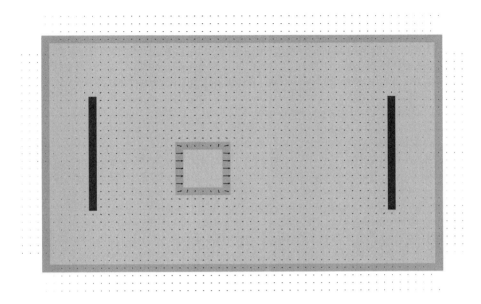

FIGURE 2.24
Calculated **E** fields in the model of Figure 2.22, but with a simple model of a nonconducting membrane enclosing a saline interior placed in the saline between the plate electrodes.

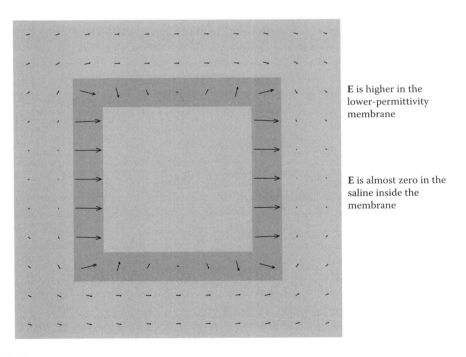

E is higher in the lower-permittivity membrane

E is almost zero in the saline inside the membrane

FIGURE 2.25
A close-up view of the **E** fields around and inside the membrane of Figure 2.24. The fields are higher in the lower-permittivity membrane and nearly zero inside it.

from the body. This section will provide a survey of the many applications of electrodes and low-frequency electromagnetic fields in medical applications.

Electrodes interact with the basic elements of the nervous system, the neurons that combine to create nerves, which transmit impulses throughout the body. A protective fatty coating called myelin insulates the nerve fibers, so that several fibers can run side by side in a bundle without interfering with each other. In this respect, nerves work much like transmission lines and are sometimes modeled this way electrically. The nerve cell uses a combination of electricity and electrochemistry. The synaptic terminals at the ends of the nerve hold neurotransmitter chemicals in membranous sacs. When the electrical signal reaches the end of the nerve, the neurotransmitters are released. The chemical neurotransmitters spread across the gap between the neurons, and at the next neuron stimulate the production of an electrical charge, which carries the nerve impulse forward into the next neuron. Thus, along the length of a neuron, the signal is purely electrical, and at the junctions, it is electrochemical. Electricity can be used to stimulate the neuron either along its length or at a junction. Receiving electrical signals must generally be done along the length of a neuron.

For all applications using electrodes at low frequencies, it is important to remember that the current passes from one electrode to another. The body region of interest should be in between the two electrodes. Practically, it is often assumed that the current passes uniformly between the electrodes, but this is rarely the case in the body because of the nonuniformity of the tissues. The current density will be greater in the high-water-content tissues (high conductivity) and lower in the low-water-content tissues. Also, the current concentrates on corners or points, so electrodes meant to provide uniform coverage over a relatively large area are generally circular or have rounded corners, as discussed in the previous section. Electrodes meant to receive from specific points (such as individual neurons) or transmit to small areas (such as implanted cardiac pacemakers) are small and pointed.

For all applications, maintaining good electrical contact between the electrode and the body is very important. Conductive gels are often used (such as in electrocardiogram measurements), and specialized materials are used to prevent buildup of nonconductive fibrous tissue (scar tissue) around implantable electrodes. The applications described below apply these principles in commonly used medical devices.

2.5.1 Electrodes for Reception

2.5.1.1 Electrophysiological Assessment

Clinical electrophysiological assessment, also called electroneuromyography (ENMG; or when only associated with a muscle, electromyography [EMG]), is used to determine the function and integrity of specific parts of the neuromuscular system. The neuromuscular junction, skeletal muscles, peripheral sensory nerve fibers, reflexes, and some central nervous system pathways can be tested using ENMG, even individual motor neurons. The more localized the test electrodes, the more detailed information that can be obtained. Surface electrodes are the least localized and least invasive. Needle electrodes are more localized and more invasive. Miniaturized needle electrodes made out of silicon are the most localized and are generally meant to be implanted in the body (and hence are the most invasive). Sometimes both stimulation and recording electrodes are used simultaneously to determine the effect of stimulating a region and determine if the signals pass through the neural system as expected. Nerve conduction studies are now very common for the diagnosis of nerve disorders such as carpal tunnel syndrome and neural injury.

Both the magnitude of the signals transmitted throughout the body and their time delay are used.

2.5.1.2 Intracellular Recording: Receiving Signals from Brain and Nerves

Laboratory studies have long been able to monitor individual nerve cells external to the body. These studies have been used extensively for the development and assessment of neurotransmitter drugs and an overall study of the nervous system. Until recently, there has been no way to perform these tests in a living organism, because the nerve cells are too small. Recently, however, a number of groups have been developing very tiny electrodes and arrays of electrodes made by etching silicon, and have been able to receive signals from individual neurons.

Although still used mainly for research, this technology is rapidly progressing toward commercial applications that focus on individual nerve stimulation more than neural recording. The same miniature electrodes are useful for both. Implanting these electrode arrays is key to their success, and this requires methods to both communicate with them (see inductive telemetry, Section 1.5, and microwave telemetry, Section 3.7) and deliver power to them (inductive telemetry). Other issues include packaging the computer, memory, and communication system right along with the electrodes and managing the localized heat that is generated by the system. Signal processing methods to interpret the received signals or to most efficiently stimulate the nerves are also receiving significant research attention.

2.5.1.3 Impedance Imaging

It would be extremely useful if impedance measurements could be used to map the impedance of the interior of the body. Such a map would provide not only morphometric information like that provided by x-ray computed tomography (CT) scans, but also additional physiological information. For example, abscesses would likely show clearly on such a map because their conductivity would be different from that of surrounding tissue, while their density, which is what is measured in an x-ray CT scan, might not be much different from surrounding tissue.

One method for mapping impedance would be to use an array of electrodes around the periphery of the body, as illustrated in Figure 2.26, and sequentially apply voltages between pairs of electrodes, measuring currents through all electrodes each time. The cross section of the body would then be divided into mathematical cells and the data used to calculate the conductivity or permittivity of each of the mathematical cells by solving a set of simultaneous equations. This is similar to some reconstruction techniques used in x-ray CT scans.

These reconstruction techniques work well with x-rays because the x-rays travel in straight lines through the body. These same reconstruction techniques do not work well with impedance imaging, however, because the currents do not travel in straight lines

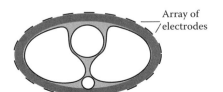
Array of electrodes

FIGURE 2.26
An array of electrodes that could be used on the surface of the body for impedance imaging.

between the electrodes through the body, as illustrated in the next section. Special techniques have been used to try to account for differing current paths, but with limited success. In the human thorax, the low-impedance path of the chest wall surrounding the high-impedance lungs shunts electrode currents around instead of through the interior, making it very difficult to get enough imaging data from the interior to provide useful images of the lungs, no matter what reconstruction techniques are used. Furthermore, the resolution (size of the mathematical cells) is limited by the number of electrodes used. An impractically large number of electrodes would be required for very small mathematical cells.

Although impedance imaging may find some practical uses, it appears unlikely to approach the resolution and usefulness of x-ray or magnetic resonance imaging methods. If impedance imaging could be made useful, it would have the advantage of being simpler and less expensive than some other imaging techniques.

2.5.1.4 Impedance Monitoring for Lung Water Content and Percent Body Fat

The idea of using EM fields for diagnosing conditions inside the human body is appealing because the distribution of internal E fields is strongly dependent on the conductivity and permittivity of the tissues, and the conductivity and permittivity of the tissues vary widely with tissue type and condition. For example, the conductivity and permittivity of muscle tissue are roughly two to twenty times greater than those of fatty tissue, depending on the frequency. With such a contrast in tissue electrical properties, it would seem that EM fields could be used to sense the presence and condition of various tissues, and thus provide diagnostic information about the interior of the body. One possible way to use this diagnostic information is with impedance measurement between two electrodes placed on the surface of the body. Impedance is a complex value and is measured by taking the ratio of the phasor voltage to phasor current. This means that both the magnitude and phase of the voltage and current must be measured. The lower the frequency, the less variation that is expected in the phase of the voltage and current. Thus, for direct current (DC), often just the resistance (magnitude of the voltage divided by magnitude of the current) is used. This simplifies the measurement considerably but loses some of the information otherwise available.

The design and placement of the electrodes are very important to the effectiveness of impedance measurement. The field distribution between the electrodes should pass through the region of interest with minimal fringing into regions not of interest. Because of the low frequencies used with electrodes, it is not possible to contain the fields to a small region when using external electrodes, so mostly very large regions of the body are considered. The basic idea is illustrated in Figure 2.27, which shows the cross section of an elementary model that has some of the gross features of the human thorax, heart, lungs, and spine. The impedance measured between electrodes A and B might be expected to reflect primarily the electrical properties of the lung, if the currents between the electrodes depended primarily on the impedances along the straight-line path between them. This could be useful, for example, in monitoring lung water content, since the impedance of a wet lung would be expected to be considerably lower than that of a dry lung. Similarly, the impedance measured between electrodes C and D would be expected to reflect primarily the electrical properties of the heart.

However, the currents between the electrodes do not depend primarily on impedances along the straight-line path between them. They tend to follow the paths of least impedances between electrodes, not necessarily straight-line paths. In Figure 2.27, the currents

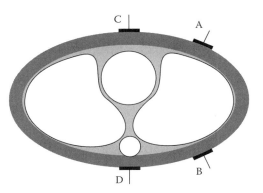

FIGURE 2.27
Cross section of a gross model of the human thorax that includes the heart, lungs, and spine.

between electrodes A and B will tend to go around the lungs through the chest wall, which has a much lower impedance than the lungs. Because only a portion of the current flows through the lung, the impedance contains less information about the lung. This effect is illustrated in Figure 2.28, which shows a simplified two-dimensional model of just a wall and an interior, with the conductivity of the interior ten times smaller than that of the wall. A current source is connected to electrodes at points A and B. As the figure shows, the currents in the wall are larger than the currents in the interior. In a similar way, the impedance between electrodes C and D contains less information about the heart because of the shunting effect of the wall.

Various techniques have been used in attempts to overcome this difficulty. One method is to put rings around the electrodes, which are called *guard electrodes*. Guard electrodes help reduce the undesirable shunt currents, but they also reduce the sensitivity of the

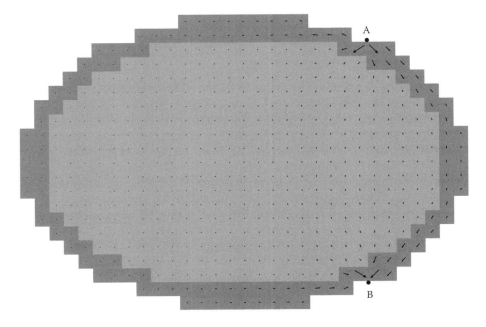

FIGURE 2.28
Calculated currents in a simplified two-dimensional model of the human thorax that includes just a wall and interior. A 100 kHz current source is connected at points A and B. Conductivity of the wall is 0.5 S/m, and of the interior is 0.05 S/m. Relative permittivity of both the wall and the interior is 3×10^4.

measurements. High impedance between the electrodes and the skin can also be a problem. Using higher frequencies helps reduce this difficulty.

Applications of impedance measurements used in research include measuring lung water content, monitoring cardiac functions, and sensing urinary bladder fullness. Variations in impedance measurements from patient to patient, depending on the patient's size and shape, present problems in developing reliable diagnostic devices based on impedance measurements. For example, the impedance measurement in one patient with a given degree of bladder fullness might be quite different from impedance measurements in another patient with the same degree of bladder fullness. Such devices might be more useful in monitoring changes in a single patient as a function of time.

One commercial application of impedance measurement is embodied in scales that measure both weight and fat content of the body. Two electrodes are placed on the scale, one for each foot, and the impedance between them is measured. Since muscle conducts current better than fat and bone and is also heavier, a combination of information on weight and impedance could theoretically provide information on percentage body fat. Unfortunately, the overall impedance also depends on the contact impedance between the foot and scale (so measurement after a bath or shower is preferred). Also, since bone and fat have very similar impedances but different weights, the relative fraction of bone must be known. Thus, the user is asked to enter weight and approximate stature (height and small, medium, or large bone structure). Since all of these parameters control the impedance of the body, they make it very difficult to get an accurate numerical evaluation of the percentage fat from individual to individual. The relative percentage over time within one individual is much easier to predict. Therefore, these scales are likely to give reasonable values for changes in percentage fat for one person, but are less likely to give accurate values for the actual percentage at any given time.

2.5.2 Electrodes for Stimulation

2.5.2.1 Cardiac Pacemakers and Defibrillators

The heart is a natural example of a critical bioelectromagnetic system. The beating of the heart is controlled by two major internal stimulation centers that are connected and usually coordinated. The sinoatrial (SA) node is located in the wall of the right atrium near the superior vena cava. The atrioventricular (AV) node is in the interatrial septum near the opening of the coronary sinus. To begin a heartbeat, the SA node creates a pulse whose potential rapidly rises from -70 mV to $+20$ mV (relative to the extracellular fluid); this pulse contracts the atria and travels to the AV node, where it undergoes a slight delay. The AV node is connected to the rest of the heart via a fibrous bundle (bundle of His) that divides into left and right branches and connects to the ventricular Purkinje tissue. The fibers that make up this bundle have been specifically modified to conduct electrical impulses to the other coronary tissues to cause them to contract and pump blood within the heart. The timing of the cardiac action potential between the various structures of the heart controls the nature and speed of the heartbeat.

The electrical signals from the heart are unavoidably conducted to other regions of the body as well, because of the conductive fluid and organs in contact with the heart. An electrocardiogram (ECG), which is used to diagnose the control signals of the heart, is a mapping of the electrical voltages produced by the heart on the surface of the body. External electrodes are placed on the body's surface to receive these electrical signals. As with any voltage, the surface potentials must be measured relative to a reference (or ground) electrode. In the simple three-lead case, the ground electrode is connected to the right leg. This means that the other two voltages that are measured will be measured relative to the right leg. When six electrodes are used, the potentials between any pair can be measured as well as the potential between each electrode and ground.

Cardiac electrical signals sometimes malfunction. During a heart attack, for instance, the heart may go into fibrillation where the electrical signals become confused and do not progress around the heart in the correct order, thus not allowing the heart to contract properly for pumping of blood. Then an external defibrillator may be used. This defibrillator uses a pair of electrodes (paddles) that are placed on the chest so that the current passing between them will also pass through as much of the heart as possible. Again, one of these electrodes is ground, and the other has a large voltage, producing a shock across the heart that attempts to restart the normal electrical activity of the heart. Defibrillators can also be implanted internal to the body. In that case, the battery pack is placed in the shoulder or chest cavity of the patient. The metal case (usually titanium) of the battery pack is left in contact with the body (i.e., not insulated with plastic), and a metal lead wire is surgically attached near an electrically sensitive point on the heart. The internal battery pack can then deliver a large enough pulse to restart the heart if needed. The two electrodes in this situation are the tip of the wire attached to the heart and the case of the battery pack. A very similar device can be used for cardiac pacing. A pacemaker typically works continuously, controlling most if not all of the heart's beats, whereas a defibrillator typically works only occasionally when needed.

2.5.2.2 *Pulsed Electromagnetic Fields*

Pulsed electromagnetic fields (PEMFs) have been used for a number of medical applications. Electrodes on the surface of the body are most often used for these applications, although internal electrodes can be used in isolated applications. Pulsed electromagnetic fields, such as those described in Section 1.8, use time-varying fields to stimulate the body for pain control, bone and tissue healing, and nerve stimulation.

Pulsed electromagnetic fields have been found to be highly effective for healing fractures and soft tissue injuries, particularly those that do not respond to ordinary healing methods. This method has been used historically. As early as 1812, passing "electric fluids" through needles inserted in the fracture gap was found to stimulate bone healing, and by the mid-1800s, this DC stimulation was considered the method of choice for slow-healing fractures. Today's bone healing PEMF systems typically use a rectangular pulse with a repetition rate of seventy-two pulses per second. The biological mechanism for the effectiveness of this treatment is not well understood, and there are various theories. It is generally accepted that this is not a thermal effect, because the power delivered is too low to increase the temperature appreciably. It is possible that the nature of the gap itself creates stronger fields in the gap than otherwise anticipated, much like the high fields produced in fat by microwave hyperthermia. Probably the most widely held belief is that pulsed fields retard the osteoclasts that destroy bone while increasing the rate of new bone formation.

Another application of pulsed electromagnetic fields is for controlling pain in the knee, back, and shoulder. Typically many electrodes are used, placed strategically around these regions. Most practitioners place the electrodes according to the concept that they are trying to gently overstimulate the nerves to use the body's natural tendency to saturate the nerve and therefore send fewer pain signals. Thus, they attempt to focus the electrodes so that the fields between them will be primarily on the nerves. This is far from an exact science. The current density is highly nonuniform between the electrodes. The bones (very prevalent in regions with pain) do not conduct as well as the surrounding muscle, nerve, and blood. The currents between the electrodes therefore concentrate in those regions. It is also likely that the direction of the current (parallel or perpendicular to the nerve) is an important consideration. In practice, most electrotherapists will simply experiment with an individual patient, and use feedback from the patient to place the electrodes in the most effective locations. This method is used extensively in both human and veterinary medicine.

Pulsed electromagnetic fields are also used for a wide variety of needleless drug delivery applications. Iontophoresis is a method to electrically force drugs across a transdermal interface using a relatively small voltage (0.1 to 10 V) across the skin boundary. This method appears not to create structural changes in the cells or the skin, but rather just creates ion pathways that a conductive fluid (drug) will follow through preexisting aqueous pathways. At present, a limited number of drugs can be delivered in this method. This method has been extended to a relatively new cancer treatment called electrochemotherapy, which has been used for a variety of cutaneous tumors, including head and neck tumors, melanomas, superficial breast cancer lesions, and so on. In this therapy, the resistance of malignant cells to penetration by certain chemotherapeutic agents is temporarily lowered by electroporation, which creates temporary pores (pathways) in the membranes of the malignant cells by the application of short DC pulses that generate electric fields of several kilovolts per centimeter. Once the cells are porated, the chemotherapeutic agents can enter the malignant cells and destroy them. Electrochemotherapy cannot only increase the efficacy of certain chemotherapeutic agents, but also can reduce side effects because malignant cells can be destroyed with much lower doses of chemotherapeutic agents than with conventional systemic chemotherapy. This method is fundamentally different from hyperthermia (heating) combined with chemotherapy, which is described in Chapter 6.

2.5.2.3 Direct Nerve Stimulation

Electrodes for neural recording were described in the previous section. These same electrodes have been used for neural stimulation as well. Deep brain stimulators for treatment of Parkinson's disease currently use a device very similar to a cardiac pacemaker. A battery pack is imbedded in the shoulder, and a wire lead is run up the neck to the brain, where an implanted electrode can stimulate the brain to relieve tremors. Future versions of this device may utilize more compact electronics directly integrated with the stimulation electrodes, thus obviating the need for the shoulder battery pack and the long lead to the brain.

Both external electrodes and invasive needle electrodes are used for stimulation of nerves and muscles for therapy and diagnostic purposes. Direct nerve stimulation has a wide array of probable applications on the horizon as well. Stimulation of the optical nerve for retinal implants and the auditory nerve for cochlear implants, as well as bypassing a damaged spinal cord, have all been demonstrated in the laboratory. Treatment of urinary, rectal, and erectile dysfunction has also been proposed using implantable neural stimulators. These devices still have many years of development before reaching mainstream application, but have shown sufficient promise to generate a large amount of research at both the academic and commercial level.

2.5.2.4 Ablation

Ablation refers to using large amounts of localized heat (burning) to treat medical conditions. Cardiac ablation uses this localized heat to destroy ill-conducting pathways in heart muscles. In coronary arteries, it is used to break up plaque and clear the arteries. This is typically done either with microwave power delivered very locally (usually with an open-ended coaxial line or a very small monopole antenna) or with a resistive wire or coil placed in contact with the plaque. Laser ablation is often used to treat endometriosis by burning or removing unwanted tissue on the surface of the uterus. All ablation applications require very precise control of the localized heating.

2.6 Fields Induced in Objects by Incident B Fields in Free Space

Figure 2.29 shows the calculated internal **E** in a simple two-dimensional model of a conducting prolate spheroid placed in a 60 Hz sinusoidally time-varying and spatially uniform **B** field in free space. The **B** field in this case is directed out of the paper (the **B** field is parallel to the minor [short] axis of the spheroid, which is also directed out of the paper). As explained in Section 1.4, the time-varying **B** field produces **E** fields that encircle the **B** field. A close-up view of the **E** fields in the spheroid is shown in Figure 2.30. When the **B** field is parallel to the major (long) axis of the spheroid, the induced **E** field pattern is as shown in Figure 2.31, which shows one cross section of the spheroid, with the **B** field directed out of the paper. A view with a finer grid is shown in Figure 2.32. The pattern is similar in each cross section of the spheroid perpendicular to the major axis. As in the other orientation of the object with respect to the **B** field, the induced **E** field encircles the **B** field. Although the rectangular mathematical cells of the model do not represent the smooth boundaries of a prolate spheroid very well, the pattern of induced **E** fields is approximately the same as that for an actual prolate spheroid.

Figure 2.33 shows the results of three-dimensional analytical calculations for the **E** fields induced in a prolate spheroid by a sinusoidally time-varying, spatially uniform **B** field. When the **B** field is parallel to the major axis of the spheroid, as shown in Figure 2.33(a), the induced **E** field in any cross section perpendicular to the major axis is given by $\omega Br/2$, where r is the distance from the spheroid major axis to the point at which the **E** field is evaluated, and ω is the radian frequency of the **B** field. This shows that the higher the

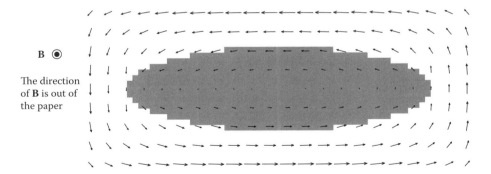

FIGURE 2.29
Calculated **E** fields in a cross section of a two-dimensional model of a prolate spheroid in free space exposed to a uniform 60 Hz **B** field perpendicular to the major axis of the spheroid (i.e., **B** is directed out of the paper). The fields were calculated on a finer grid and displayed on a coarser grid to show more clearly the overall field pattern.

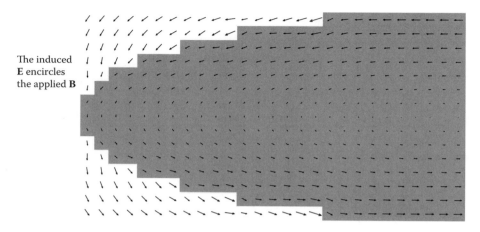

FIGURE 2.30
The **E** field pattern in the left half of the spheroid in Figure 2.29, shown as calculated on the finer grid.

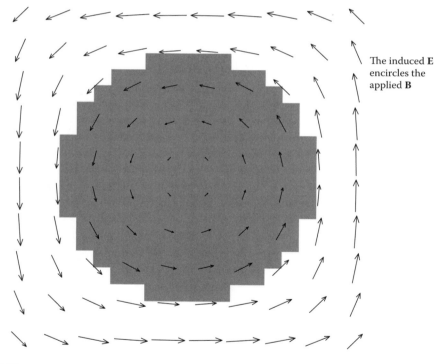

FIGURE 2.31
Calculated **E** fields in a cross section of a two-dimensional model of a prolate spheroid in free space exposed to a uniform 60 Hz **B** field perpendicular to the minor axis of the spheroid (i.e., **B** is directed out of the paper). The fields were calculated on a finer grid and displayed on a coarser grid to show more clearly the overall field pattern. The **E** encircles the applied **B** and is stronger near the outside of the object.

radian frequency ω, the more electric field that is produced. Also, the farther from the axis, the greater the electric field. The maximum **E** field induced by a **B** field of 1 mT at a frequency of 60 Hz (ω = 2π × 60) is 25.8 mV/m, located at the outer surface of the spheroid.

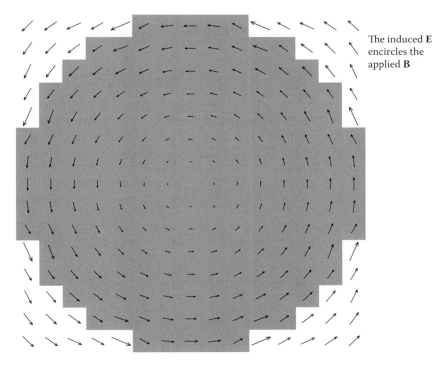

The induced **E** encircles the applied **B**

FIGURE 2.32
A close-up view of the fields in Figure 2.31, showing the fields calculated on the finer grid.

When the **B** field is parallel to the minor axis of the spheroid (Figure 2.33(b)), the maximum induced **E** field, located at the outer surface of the spheroid, is almost twice that value. The difference can be explained in terms of the cross-sectional area intercepted by the **B** field in each case. In Figure 2.33(a), the cross-sectional area intercepted by **B** is considerably less than that intercepted by the **B** field in Figure 2.33(b). This is an illustration of a general behavior: the **E** fields induced in a body by a spatially uniform **B** field are generally greater when the cross-sectional area intercepted by **B** is greater, and are found near the outer periphery of the body.

The **E** fields induced in a coarse two-dimensional model of an animal in free space exposed to a 60 Hz, spatially uniform **B** field are shown in Figure 2.34. Again, the **E** fields tend to circle around the applied **B** field, which is directed out of the paper. They are generally larger in the air surrounding the conducting tissue of the model than they are in the tissue itself. The **E** fields also tend to be small near the center of the system and larger around the outside, as in the spheroidal models. Figure 2.35 shows the **E** fields inside the model only (plotted to a different scale), where the circulating pattern is more obvious.

It is interesting to note that the **E** fields tend to circulate around the center of the trunk, but also to a lesser extent around the center of the head and the center of the legs. The circulation around the center of the left leg is shown more clearly in Figure 2.36, which shows just the left leg of Figure 2.34, but still attached to the whole animal. Figure 2.37 shows the **E** fields in a leg that has been detached from the rest of the animal but exposed to the same **B** field, plotted to the same scale as in Figure 2.36. Although the fields at the very top of the detached leg are different from those of the attached leg, the fields at the bottom of the detached leg differ from those in the attached leg by less than one-half of 1%. This

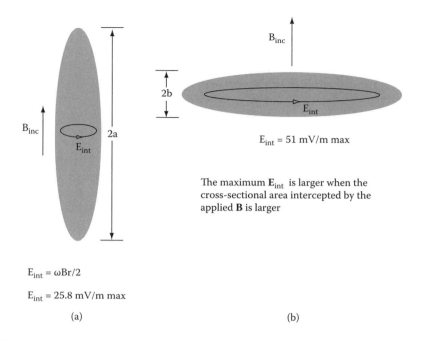

$$E_{int} = 51 \text{ mV/m max}$$

The maximum E_{int} is larger when the cross-sectional area intercepted by the applied **B** is larger

$$E_{int} = \omega B r / 2$$

$$E_{int} = 25.8 \text{ mV/m max}$$

(a)   (b)

FIGURE 2.33
Comparison of the internal fields in two spheroids, (a) with the incident **B** parallel to the long axis of the spheroid, and (b) with the incident **B** perpendicular to the long axis of the spheroid. In both cases the incident **B** is 1 mT, the conductivity is 0.067 S/m, the frequency is 60 Hz, a = 0.875 m, and b = 0.138 m. The internal **E** fields were calculated in three dimensions using a long-wavelength approximation to Maxwell's equations. The **E** fields are stronger when the cross-sectional area that **B** passes through is larger (case b).

comparison indicates that the fields tend to circulate around the center of the attached leg as though it were a separate entity, and only in the region where the leg is attached are the fields significantly different from those of a detached leg. This effect becomes more pronounced as the leg becomes longer and thinner, and less pronounced as the leg becomes shorter and fatter (and blends more into the body as a whole).

Figure 2.38 shows a close-up of just the **E** fields in the attached head and neck of the model of Figure 2.34. The pattern very clearly shows how the fields are circulating around the center of the head, almost as if it were detached. Only the fields near the neck are significantly different from those of a detached head. This effect becomes more pronounced as the area of the appendage becomes a greater portion of the entire area.

POWER LINES AND PEOPLE

Whether or not living under or near power lines causes cancer (particularly leukemia and some types of brain tumors) continues to be a question for international debate and research. The late 1980s brought a number of epidemiological studies from the United States and Europe that showed statistical links between childhood leukemia and proximity to power lines. Later European studies showed weaker or nonexistent links, particularly when normalizing for the distance from roadways (car exhaust is a known carcinogen). At power line frequencies *(continued on next page)*

(60 Hz in the United States and 50 Hz elsewhere) the electric and magnetic fields are decoupled. They do not generate each other and can be evaluated independently. The magnetic field from a power line encircles the line (according to the right-hand rule described in Section 1.3). If a person is standing near the power line, this can be evaluated as a frontally incident magnetic field. This field will generate circulating electric fields within the body. However, biological effects from the magnetic field have generally not been implicated in the debate, and attention has focused on the electric field and the associated currents within the body.

The electric field goes from the power line to the ground and is vertically polarized with respect to a standing person under the power line. This field enters the body through the head and shoulders, passes through the torso, and exits through the feet and legs, as shown in Figure 2.39. If the person is isolated from the ground (wearing tennis shoes and standing on a dry surface, for instance), the field tends to exit uniformly from the legs, as illustrated in Figure 2.39(a), but if the person is grounded (wearing leather-soled shoes and standing in wet grass), the current passes out mainly through the bottoms of the feet, as shown in Figure 2.39(b).

Using the finite-difference time-domain method described in Chapter 5, the peak current density in the body is found to be in the ankle and knee, as indicated in Figure 2.39(c). This is expected, because both of these regions are relatively bony (bone does not conduct much) with very little conductive material in their cross section. The current passing down through these regions is concentrated in the relatively small regions of surrounding muscle and fat, giving large current density. A less anticipated result was that there are regions of large current density in the torso also. This is because the lungs are also not very conductive, and the current must flow through the outer region of the torso, mainly in the muscle regions. The muscles of the back, directly behind the lungs, have several regions with large current density because of this concentration of the current.

Epidemiological and bioeffect research continues today. The bioeffect research often uses laboratory animals instead of humans. The current density in these animals is obviously going to be very different than in a human because of their difference in size and orientation. For a 1 μT, 60 Hz magnetic field, for instance, a human will have a calculated average current density of 1.3 to 1.9 μA/m², while a rat will have 0.3 μA/m² and a mouse 0.12 μA/m². Under these conditions, the maximum calculated current density for the human is 8 μA/m², the rat 1.3 μA/m², and the mouse 0.4 μA/m². This research relies on dosimetry (Chapter 5) to determine the relative doses for different exposure conditions and different animals.

2.7 E Field Patterns for *In Vitro* Applied B Fields

In typical laboratory experiments, a biological sample is exposed to a **B** field produced by currents in a coil or a combination of coils, with the **B** field being approximately spatially uniform over the region of the biological sample. In the calculations described below, we have approximated the laboratory **B** fields with spatially uniform **B** fields, which is often a satisfactory approximation.

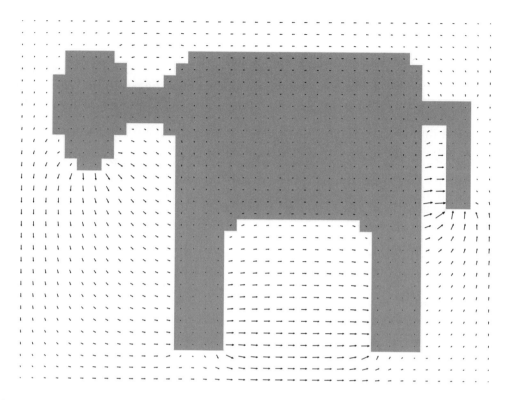

FIGURE 2.34

Calculated **E** fields in a two-dimensional coarse model of an animal in free space exposed to a uniform 60 Hz **B** field in a direction out of the paper. The conductivity of the animal is 0.6 S/m. Even though the applied **B** field is uniform, the **E** field distribution within the animal is clearly nonuniform. The animal significantly perturbs the fields both inside and outside of the animal.

Calculated **E** fields induced by a spatially uniform, sinusoidally time-varying 60 Hz applied **B** field in a two-dimensional model of a saline solution in a round container with nonconducting walls are shown in Figure 2.40. Consistent with the discussions related to previous examples, the **E** field circulates around the applied **B** field. This is shown more clearly in Figure 2.41, which displays the **E** fields on a coarser grid, so that the arrows are longer, and the overall pattern is more apparent. On this coarser grid, however, the **E** fields in the wall of the container are no longer identifiable.

The square corners in the patterns occur because of the rectangular mathematical grid used to make the calculations. (**E** fields concentrate on corners, even when the corners are artificially induced by the numerical model. We normally call these numerical modeling artifacts stair-stepping errors and ignore them, but see below.) A more realistic pattern would be obtained by using a very large grid with many more, smaller mathematical cells. This would require more computer memory and longer calculation time, but the resulting pattern would be more circularly symmetric.

In some simulations, such as modeling curved metal surfaces, the stair-stepping effects can never quite be eliminated, and extreme care must be taken when evaluating the fields on or very near the stair-stepped surface. For instance, if SAR or current density were evaluated on the surface of a metal coil used for nuclear magnetic resonance (NMR) or magnetic resonance imaging (MRI), the stair stepping of the coil could make these values

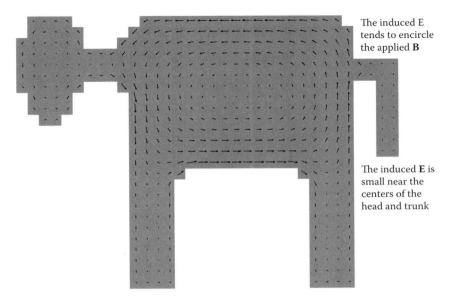

The induced E tends to encircle the applied **B**

The induced **E** is small near the centers of the head and trunk

FIGURE 2.35
The same model as in Figure 2.34, but showing only the **E** fields inside the animal plotted to a different scale to show the internal fields more clearly. As with a simpler object (Figures 2.30 and 2.32, for instance), the fields tend to circulate around the applied **B** field, although this time several individual regions of circulation are seen (body, head, legs). They are also stronger near the outside of the animal.

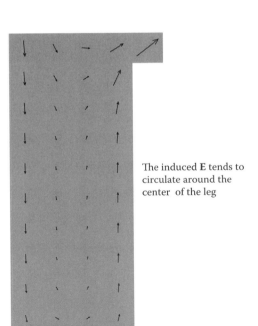

FIGURE 2.36
A view of just the left leg attached to the model of Figure 2.34, and the **E** fields inside it, showing its individual region of circulation.

The induced **E** tends to circulate around the center of the leg

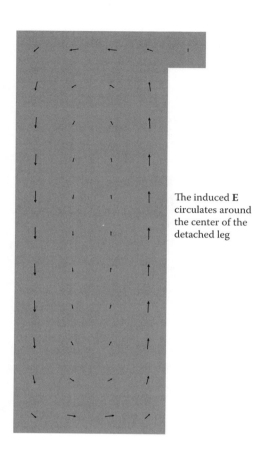

FIGURE 2.37
The **E** fields inside a detached leg of the model of Figure 2.34, plotted to the same scale as those in the attached leg in Figure 2.36.

The induced **E** circulates around the center of the detached leg

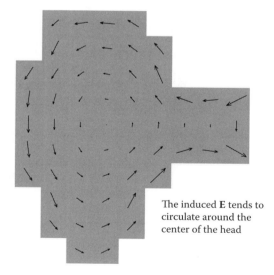

The induced **E** tends to circulate around the center of the head

FIGURE 2.38
A view of just the **E** fields in the attached head and neck of the model of Figure 2.34, showing its individual region of circulation.

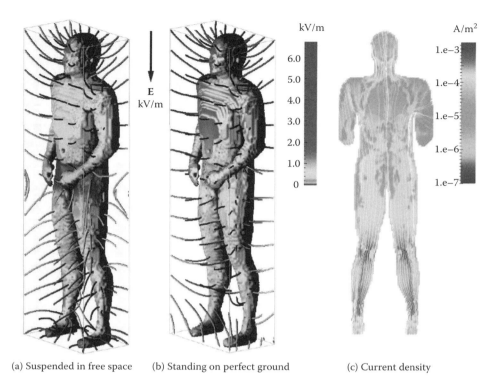

(a) Suspended in free space   (b) Standing on perfect ground   (c) Current density

FIGURE 2.39
Please see color insert following page 146. The electric field vectors intersecting a human body standing in a uniform 1 kV/m vertical electric field at 60 Hz for the case of (a) insulated feet and (b) grounded feet. The current density in a vertical cross section of the body is shown in (c). (From Stuchly, M., and Dawson, T., *Proc. IEEE*, 88, 643–64, 2000. © 2000 IEEE. With permission.)

appear much larger than they really would be on a smooth coil and lead to erroneous conclusions. Special care must be taken when calculating peak values, as these will commonly show up (erroneously) in stair-stepped locations. When using numerical simulations to calculate fields, it is wise to look (visually) at the spatial distributions of the fields near the peak values to ensure that they make sense before relying on them. When fields are averaged over several points (as in the case of calculating 1 or 10 g averaged SAR, for instance), this problem tends to be reduced, but a visual check of even these averaged peak values is a good idea.

An interesting question arises about the induced **E** field pattern if a conducting object were placed in the saline solution of Figure 2.40, such as the round object shown in Figure 2.42. You might expect that the basic pattern of Figure 2.40 would be somewhat modified by the presence of the object, but that the pattern in the object would be similar to the pattern in the saline at that position before the object was placed there. To the contrary, however, the close-up view of Figure 2.43 shows that the **E** field inside the conducting object tends to circulate around the object's center, not the center of the dish, as it does in the absence of the object. The center of circulation inside the object is slightly offset, but close to, the center of the object. The fields inside the object are smaller than in the surrounding saline because the conductivity of the object is 8 S/m, while the conductivity of the saline is 0.6 S/m. The presence of the object also changes the pattern in the saline.

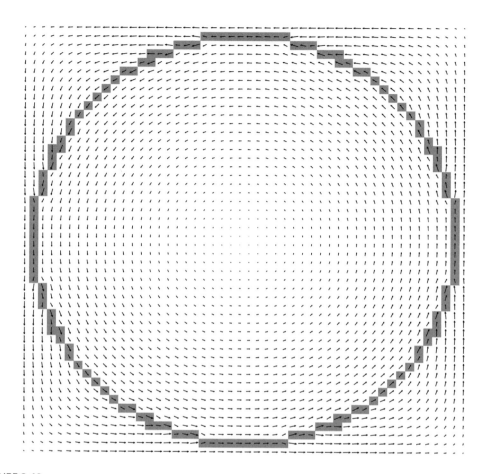

FIGURE 2.40
Calculated **E** fields in a two-dimensional model of a saline solution with a conductivity of 0.6 S/m in a round nonconducting dish exposed to a uniform 60 Hz **B** field in a direction out of the paper.

Plots of the *currents* in the saline and in the object, as shown in Figures 2.44 and 2.45, give additional insight into the effects caused by the presence of the object. The currents in the object are considerably stronger than in the saline, as shown most clearly in Figure 2.45. Thus, while the higher conductivity of the object causes the **E** fields to be weaker in the object than in the surrounding saline, it also causes the induced currents ($\tilde{J}_c = \sigma_c \tilde{E}$; see Sections 1.6 and 1.14) to be higher in the object than in the surrounding saline. As explained in connection with Figures 1.13 and 1.14, the current pattern inside the object can be thought of as consisting of a global component circulating around the center of the dish, and a local component circulating around the center of the object. Because the global component is the much weaker of the two, the combination of the two results in a circulating pattern in the object that is slightly offset from the center of the object.

To illustrate this effect further, we placed the object at the center of the dish, as shown in Figure 2.46. With the object at the center, the global currents and the local currents are one and the same because they both circulate around the same center. Thus, Figure 2.46 shows what the local currents are. Figure 2.47 shows a closer view of these currents. It is apparent

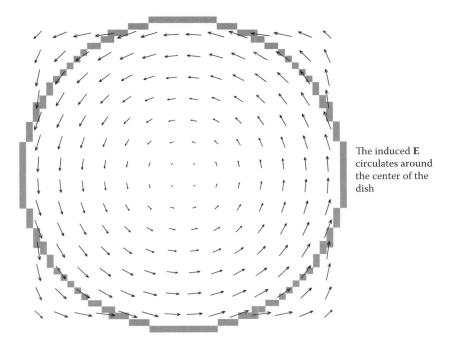

The induced **E** circulates around the center of the dish

FIGURE 2.41
The **E** fields of Figure 2.40 displayed on a coarser grid to show the field patterns more clearly. Because of the coarser grid, the fields in the wall of the container are not shown specifically. The fields tend to circulate around the applied **B** field (and thus the dish) and are stronger toward the outside of the dish.

that the local currents in Figures 2.45 and 2.47 are nearly the same, as expected, because the conductivity of the object is much greater than that of the saline.

The behavior illustrated in Figures 2.44 and 2.46 is characteristic of a conducting object placed in a conducting solution exposed to a spatially uniform sinusoidal **B** field. The higher the conductivity of the object compared to the solution, the stronger the local component compared to the global component, which makes the local current pattern less dependent upon the exact position of the object within the global pattern. Understanding of this behavior is important in designing experiments in which **B** fields are used to expose biological preparations.

A contrasting behavior occurs if the object placed in the solution is nonconducting. The induced **E** field pattern for this case is shown in Figure 2.48. The **E** fields inside the object are now stronger than those in the surrounding saline. The insulating properties of the object also cause the induced **E** fields to be stronger in the narrow region between the object and the container wall. The plot of the currents shown in Figure 2.49 helps us understand this. The current in the object is negligible because it consists only of displacement current, which is very small because the frequency is so low. Consequently, the current is forced to flow around the object, causing current concentration in the narrow region between the object and the container wall.

What would happen if metal plate electrodes (not connected to any source) were added to a container of conducting solution exposed to a uniform **B** field? To consider this question, we first show in Figure 2.50 the **E** fields induced in a square container of saline exposed to a uniform **B** field with no electrodes. The different pattern that results when metal electrodes are added is shown in Figure 2.51, with a close-up view of the upper half shown in

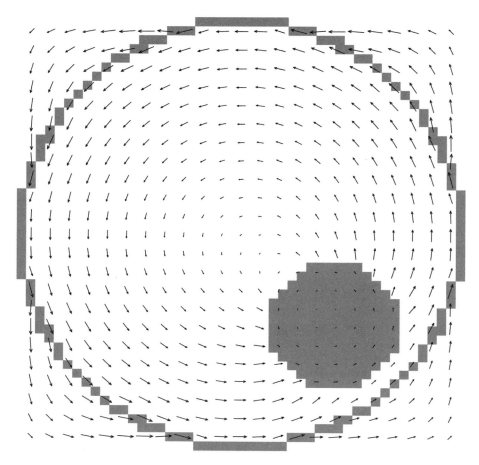

FIGURE 2.42
Calculated **E** fields in the model of Figure 2.40, to which a round object with conductivity of 8 S/m has been added. The fields were calculated on a finer grid and displayed on a coarser grid. The conducting object has lower fields, and perturbs the field pattern both inside and outside of itself.

Figure 2.52. The electrodes straighten out the **E** field lines so that in between the electrodes they no longer are circular around the center of the dish. The **E** field pattern is similar in some respects to the pattern produced by a current source connected between plate electrodes, as shown in Figure 2.22. There is one major difference in the patterns, however. The **E** fields in Figure 2.51 vary approximately linearly with the distance from the horizontal centerline of the dish; note that they are zero along the centerline, and even change direction between the top and bottom halves of the electrodes. Except near the ends of the electrodes, they are relatively uniform, moving from left to right between the electrodes.

What would the induced **E** fields be in a nonconducting membrane model placed in the configuration of Figure 2.51? The results are shown in Figure 2.53, with a close-up view of the membrane in Figure 2.54. The results are very similar to those of Figure 2.25, in which the membrane was placed between electrodes excited by a current source.

From the results of this two-dimensional simulation, we might conjecture that metal electrodes in saline exposed to a uniform **B** field could be a good system for *in vitro* exposures to **E** fields, if the sample could be placed in a relatively small region up (or down) from

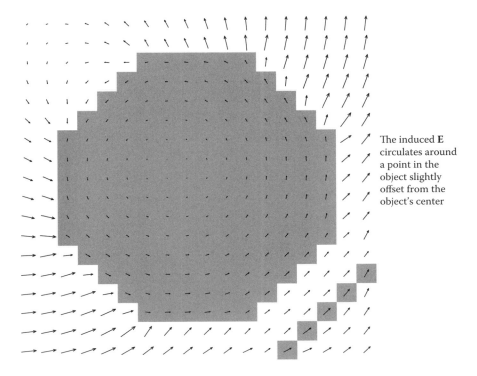

The induced **E** circulates around a point in the object slightly offset from the object's center

FIGURE 2.43
A close-up view of the field pattern in the region of the object in Figure 2.42, displayed on the finer grid. The fields tend to circulate within the object as well as outside of it.

the centerline of the dish. This system has the advantage that it does not require external connections to the metal electrodes, but it does have several disadvantages. The induced **E** field is much less spatially uniform in this system than it is in the system excited by a current source. Furthermore, the magnitude of the **E** fields would be expected to be much smaller for practical values of **B** field excitation than for practical values of current-source excitation. Careful design would be required to determine whether this kind of **B** field excitation could be practical.

Once again, these examples illustrate the effect that measurement probes can have on the field distribution by changing it in an attempt to measure it—if proper care is not taken to ensure that their impact is minimal. Any object, metallic or dielectric, placed in an EM field may change the behavior of the field. The hands of a person holding a petri dish, nearby petri dishes, stirring rods or beads, or the metal walls of a test chamber all have the ability to change the field distribution in the original dish.

2.8 Measurement of Low-Frequency Electric and Magnetic Fields

At low frequencies, the electric and magnetic fields are uncoupled, which means they do not generate each other. Thus, they can and should be measured separately, and their biological effects are generally considered independently. Low-frequency fields have such long wavelengths that phase shift is often negligible.

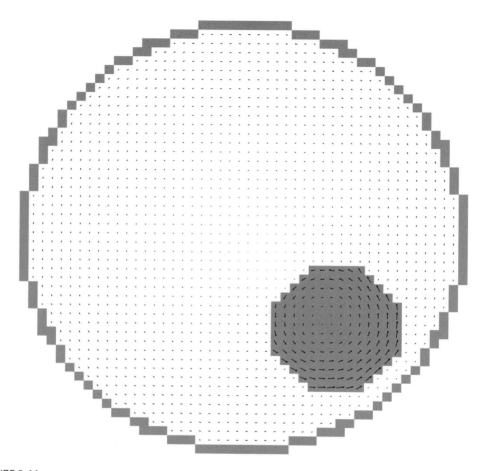

FIGURE 2.44
A plot of the current (conduction plus displacement) in the model of Figure 2.42. The displacement current is
negligible, as indicated by the absence of current in the wall of the container.

Before we discuss the specifics of how electric and magnetic fields are measured, it is
important to understand the difference between *analog* and *digital* electrical signals. Let us
consider voltage, for example. Voltage is an analog value, which means it can take on any
value (for example, 1.2536 … V) within its possible range. Computers store and process
digital values, however. A digital value is specified with a series of bits, which are either
1 or 0. (Computers store these as either "on" or "off" electrical signals.) A 2-bit system will
have four possible measurements (defined by 0-0, 0-1, 1-0, 1-1). So if a 2-bit system measures
voltages from 0 to 3 V, 0-0 would represent 0 V, 0-1 would represent 1 V, 1-0 would represent 2 V, and 1-1 would represent 3 V.

The challenge when using digital representation is that any value between two digitally
defined voltages would be rounded either up or down. There are no intermediate values
in a digital signal, but of course, there are an infinite number of intermediate values in an
analog signal. The more bits in a digital signal, the more intermediate values it can represent. For n bits, there are 2n possible values. Analog signals are almost always converted to
digital signals today using analog-to-digital (A-to-D or AD) converters that are commonly

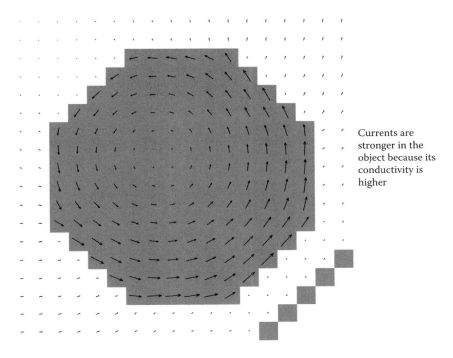

Currents are stronger in the object because its conductivity is higher

FIGURE 2.45
A close-up view of the currents in the region around the object in Figure 2.44. The currents tend to circulate within the object, but are off-center because of the nonuniform **E** fields incident on the object (because of its off-center placement within the dish).

available in 8-, 10-, 12-, 16-, and 32-bit options. The number of bits in a measurement defines the accuracy with which it can be sampled, stored, and processed by a computer. The more bits, the more accuracy, and also the more computer memory and processing power required to store and use them.

Virtually all signals we experience today are digitally stored—audio and photo images are two signals experienced regularly. The pixilation in photos is a function of the sampling that converts the analog image of the picture to a digital signal the computer can manipulate. The limited number of color possibilities available in a digital photograph is a function of the number of bits used by that camera. This is easiest to see when you are editing a photo and choose from a limited color palette. The more bits defining the color in the picture, the smoother and more refined the picture appears. And the more pixels and bits, the more memory and processing power required. Higher-megapixel cameras use more pixels and, as a result, typically take clearer pictures with better color resolution, but require larger memory cards.

A low-frequency electric field originally presents itself as an analog signal and is typically measured with two electrodes, placed at either end of the electric field path. The voltage between the electrodes is measured with standard low-frequency electrical recording equipment. Consider, for example, an electrocardiogram (ECG) signal. The heart produces a rhythmic electrical signal (voltage) within the body. This internal voltage passes through the conductive body tissues and spreads out onto the surface of the body. This external voltage can be measured with several electrodes placed on the surface of the body. To make better electrical contact (and therefore receive higher voltage) between the electrode

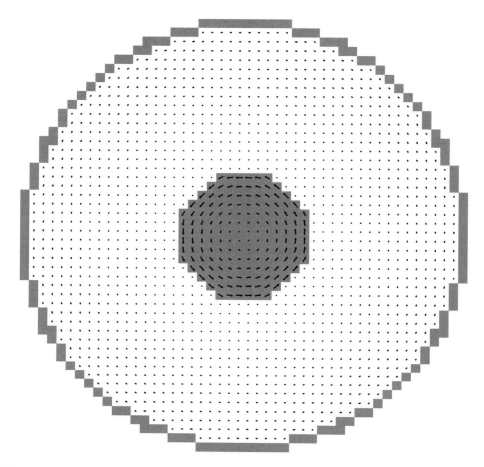

FIGURE 2.46
The current patterns when the object in Figure 2.42 is placed at the center of the dish. The currents are now centered as they circulate within the object.

and the body, conductive gel is used to lower the contact resistance. One electrode (typically placed on the lower leg, where the cardiac-induced voltage is very small) serves as the ground or reference electrode. Voltages are always measured relative to an electrical ground, but there is no such thing as a perfect ground or absolute zero volts. So we measure all electrical signals relative to some location, which we call ground and treat as if it is zero volts. For the ECG, each of the other (nonground) electrodes (typically two, three, or six) is measured one at a time relative to the ground electrode.

Since the voltage on the surface of the body is analog, the voltages picked up by the electrodes are also analog. In older ECG machines, this analog voltage was converted to a line on paper. The writing stylus was moved up and down, controlled by the magnitude of the voltage, while the paper was slowly pulled across a writing surface below the stylus. The ECG generated a visually drawn signal that a practitioner could evaluate by sight. Today the ECG voltage is converted to a digital signal and stored in a computer. It can still be plotted out in much the same as was the older stylus-based ECG machine, except that now it has a finite number of possible values, based on how many bits are used to store

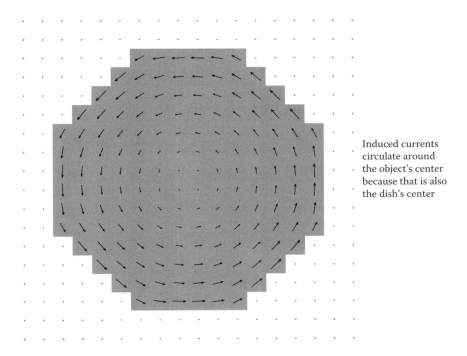

Induced currents
circulate around
the object's center
because that is also
the dish's center

FIGURE 2.47
A closer view of the currents in the object in Figure 2.46. As expected, the currents are stronger on the outer regions of the object.

the signal. This is not at all obvious to the user, because enough bits are used that the lines later plotted out appear smooth and regular.

Another important aspect of signal measurement is that the sampling rate of the time-varying cardiac signal needs to be fast enough that we do not notice any missing pieces of the signal. This is equivalent to pixels in an image. The faster the signal varies, the faster the sampling needs to be. As with all signals, the big advantage of digital signals is that a computer can be used to process them, as in the case of ECG signals where the computer automatically calculates the pulse rate and, in many cases, identifies common cardiac abnormalities.

Low-frequency magnetic fields are generally measured with loops, as described in Section 1.3. The magnetic field passing through the loop creates an electric current on the loop. When a small resistor is placed in the loop, the voltage across this resistor can be measured. This turns the magnetic field measurement into a voltage measurement also, and the same processing as for the electric field applies.

Low-frequency magnetic fields over very large areas can also be measured by measuring the current between two points and assuming the current was caused by the magnetic field. This is often how geomagnetic fields are measured on the surface of the earth. It is also common to measure the current in a wire and calculate the magnetic field that will be produced by the wire, rather than measuring the magnetic field directly.

Low-frequency fields are also often measured *differentially*, which is particularly useful for very small fields (which are often found in bioelectromagnetics). This means that two (usually very similar) fields are measured simultaneously and compared. Analog

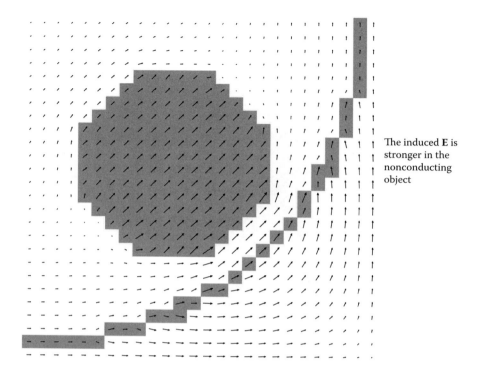

The induced **E** is stronger in the nonconducting object

FIGURE 2.48
The **E** fields in the model of Figure 2.42, but with the conducting object changed to a nonconducting dielectric object. As with other dielectric objects, the field inside the object is enhanced, and circulation is not seen. The object perturbs the fields inside of it as well as outside of it.

electronics such as differential amplifiers are often used for this purpose, particularly when the two signals are very small. If we had first converted the very tiny signals from analog to digital, a lot of information could be lost. If we had amplified each signal individually, extra noise could be added. Subtracting them first and then amplifying them in an analog fashion before digitizing gives the best possible response for small signal measurements.

Oscilloscopes are also often used at low frequencies in order to see the time-domain character of the signal. An oscilloscope has a screen that shows the signal as a function of time. This can be used to recognize unwanted or unexpected noise or jumps in the signal and is often used to debug troublesome circuits.

Low-frequency impedance measurements are also used extensively. These are generally a combination of resistance and capacitance measurements, representing the real and imaginary parts of the impedance. Inductance measurements are also used, although less commonly. Resistance is by far the most common measurement. When making low-frequency measurements the connection resistance (such as between an electrode and the skin) is quite important. Conductive gels, glues, fluids, and so on are often used to minimize contact resistance. The types of wires that are used and their lengths and orientations are usually not very important.

Another common source of error when measuring low-frequency fields, voltages, or currents is the presence of *galvanic potential*. When two dissimilar metals are put together and immersed in an electrolyte (a conductive fluid like battery acid or lemon juice), a battery

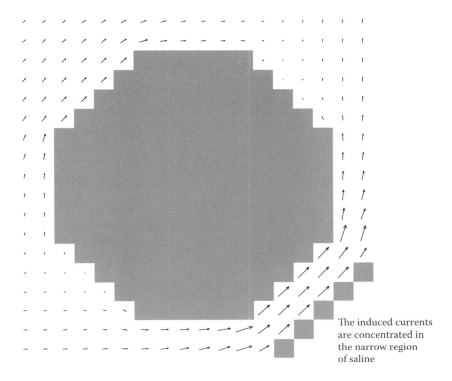

The induced currents are concentrated in the narrow region of saline

FIGURE 2.49
The current (conduction plus displacement) in the model of Figure 2.48. The displacement current is too small to be visible in the nonconducting object and the nonconducting wall of the container. The current is concentrated between the object and the wall of the dish.

is produced. The more electrically different the two metals are, and the stronger the acid, the stronger the battery. For example, a very weak battery can be produced by putting a (steel) nail and a copper wire into a potato! A similar effect, corrosion potential, is seen when similar metals are placed in two different electrolytes. For instance, moist soil is a weak electrolyte. If two (steel) screwdrivers are stuck in the ground with part of the metal in the dirt and part in the air, the metals are the same but the electrolytes (dirt and air) are different. A weak battery is produced in this case also. (Galvanic or corrosion potential is commonly seen in agricultural applications such as dairies.) The voltage from these galvanic/corrosion potential batteries can be measured using a common voltmeter. In addition, problems with *stray voltage* may be encountered. Stray voltage is any voltage that is not supposed to be in that location, and can arise from faulty or poorly grounded equipment.

The earth's geomagnetic field is also a source of potential measurement error when measuring low-strength, low-frequency fields. The magnetic field picked up by a compass, for instance, will also be picked up by magnetic field measuring equipment if it is sensitive enough. The field depends on where on the surface of the earth the measurements are taken, the time of day, and seasonal/daily variations such as solar activity.

Another source of error in low-frequency measurements is noise in the environment or in the measuring equipment itself. Many low-frequency measurements also show some high-frequency components, which can only be seen with monitoring equipment intended for use at these higher frequencies. High-frequency noise sources include, for example,

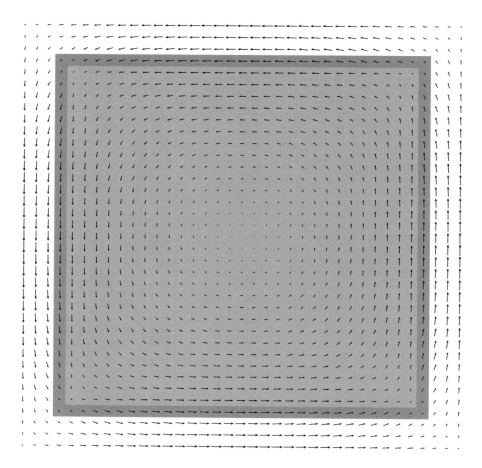

FIGURE 2.50
E field induced in a square nonconducting container of saline (no metal plates introduced yet) having a conductivity of 0.6 S/m exposed to a 60 Hz spatially uniform sinusoidal **B** field.

radio, television, and cellular telephones, and the switching of nearby large equipment (or even rather distant noises carried along a power transmission line). Many of these sources are limited to specific frequency bands and their associated harmonics, but the noise may also be a broadband signal. White noise has uniform magnitude at all frequencies. Measuring equipment always has a very small amount of white noise caused by the electronics themselves. This noise is small, and the manufacturer's specifications will usually tell you what this noise level is. Your measurements are only accurate when they are above the noise level.

2.9 Summary

Low-frequency fields are those whose wavelength is much larger than the object of interest. For human applications, this is typically from DC (zero frequency) up to the high kilohertz frequency range. Electric and magnetic fields are decoupled in this region, which allows

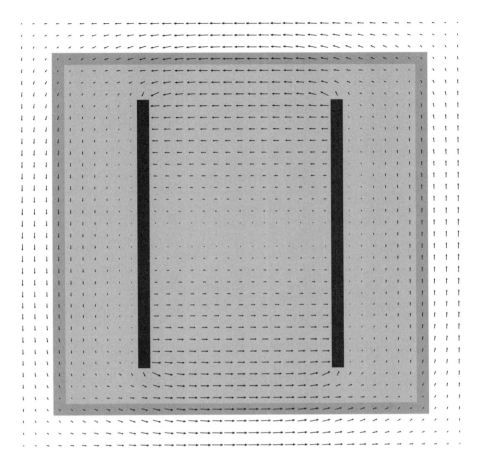

FIGURE 2.51
E fields induced in the same configuration as in Figure 2.50, but with metal plate electrodes added. The plates are not connected by any wires.

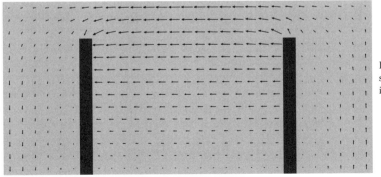

Plate electrodes straighten the induced **E**

FIGURE 2.52
A close-up view of the upper half of the saline in Figure 2.51.

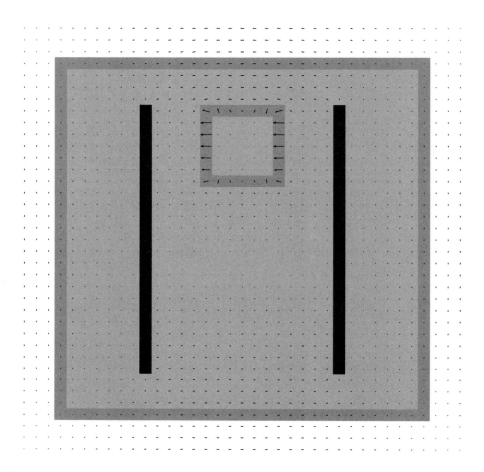

FIGURE 2.53
E fields when a nonconducting membrane filled with saline is added to the configuration of Figure 2.51.

for independent development using either electric or magnetic field sources. Four major types of applications were discussed in this chapter. Pulsed fields, such as lightning and pulsed electromagnetic fields for medical applications, deposit a large amount of power over a small area in a very short amount of time. Thus, they can be hazardous (as in the case of lightning or electrical shock), or they can provide medical benefit (such as in electrophysiology). Electrode applications for direct stimulation or direct reception of neural signals include applications such as cardiac defibrillation and pacing, cardiac ablation, and individual nerve stimulation and recording. Impedance imaging applications were also discussed, although these tend to be less effective than the imaging modalities that will be discussed in the next chapter, on mid-frequency fields. Finally, the bioelectromagnetic effects of power lines and other 50 or 60 Hz devices were discussed.

Throughout this wide variety of applications, some common concepts were seen. First, low-frequency fields follow the paths of least resistance as they propagate from high- to low-impedance regions. Thus, placing two electrodes on opposite sides of the body will force the current to pass from one electrode to the other, but it will tend to concentrate it in the conductive regions (high-water-content tissues) between them. Second, low-frequency fields cannot be easily focused, making imaging difficult. But they do focus themselves

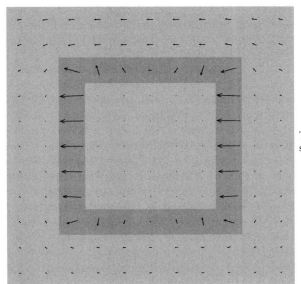

The induced **E** pattern is similar to that in Fig. 2.26

FIGURE 2.54
A close-up view of the **E** fields in and around the membrane of Figure 2.53.

on any metal corner or metal point. Thus, electrodes used to propagate fields through relatively large regions of the body (for external cardiac defibrillators, for instance) are generally rounded, whereas those used to excite or record from individual neurons are as thin and pointed as possible. Low-frequency fields are endemic to our environment and are endogenous in our bodies. Their current and potential medical applications continue to expand, as will be seen in Chapter 6.

3

EM Behavior When the Wavelength Is About the Same Size as the Object

3.1 Introduction

As explained in Chapter 1, the characteristics of EM fields and their interaction with objects are a strong function of the ratio of the wavelength of the EM fields to object size. The purpose of the present chapter is to describe and discuss the details of EM field behavior and their interactions when the wavelength is about the same size as the object. According to Figure 1.31, the free-space wavelength is 300 m when the frequency is 1 MHz, and is 0.3 mm when the frequency is 1 THz (1×10^{12} Hz). Consequently, the frequency range addressed in this chapter is approximately between these two frequencies; it is the range designated as microwave theory in Figure 1.31.

In this region of the EM spectrum, propagation effects dominate and EM field behavior is most often described in terms of wave functions. The wave functions are usually described in terms of **H** instead of **B** in this range because in the mathematical solution to the equations it is more convenient. **E** and **H** are strongly coupled together in this region of the spectrum; one cannot exist without the other. Changes in one affect (and in fact create) the other. The concept of voltage does not generally apply in this range, as it does in the low-frequency range described in Chapter 2, but voltage can be defined in special cases in this range (see Section 3.5.1). Because this chapter deals primarily with waves, their properties, and their interactions with objects, you may wish to review Section 1.11 before reading on.

At frequencies below the region considered in this chapter, EM fields can generally be treated in terms of voltages and currents. At frequencies above this region, they act more like rays. But in this middle-frequency range, they act like waves. Effects such as resonance and oscillation are often seen in this frequency range. This mid-frequency band is used extensively for communication, enabling telemetry for implantable devices, for example. External communication devices such as cellular telephones operate in this frequency range, so evaluation of their compliance with safety guidelines is important. This is because at certain frequencies, objects such as the human body absorb proportionally more power; the safety guidelines discussed in Section 5.8 reflect this fact. Some imaging is done in this frequency band, because the waves can be focused or otherwise evaluated with better spatial localization than at low frequencies. Heating is also important in this frequency range. (Hyperthermia applications are discussed more extensively in Chapter 6.)

This mid-frequency range is probably the most complex region for calculation of the expected fields because of their wave-like behavior. Waves reflect, transmit, refract, add constructively and destructively, and attenuate as they propagate through and around the body. Thus, detailed simulations such as those described in Chapter 5 are usually needed at these frequencies. In spite of this difficulty, this frequency range is home to a diverse array of applications with many more on the horizon.

3.2 Waves in Lossless Media

In this section we discuss wave properties in terms of two simple waves: spherical waves and planewaves. Both of these waves are mathematical idealizations. Neither spherical waves nor planewaves exist in perfect form physically, but they are extremely useful for conceptual understanding, and they can approximate real physical waves. Spherical waves represent what happens near a very small source, when the wave propagates away from the source in a spherical pattern. Planewaves represent what happens very far from almost any source, when the wavefronts (which perhaps started out as spherical waves) are now nearly planar. Later sections will also discuss ways in which waves that are neither planar nor spherical interact with the body.

The discussion in the following two subsections is restricted to waves propagating in a lossless material. Lossless material is material in which the effective conductivity (see Section 1.4) is zero, and no power is absorbed. Wave effects at interfaces between different materials are discussed in Section 3.3, and waves in lossy material are discussed in Section 3.4.

3.2.1 Spherical Waves

Spherical waves are perhaps the simplest kind of waves to understand. A spherical EM wave consists of an **E** field and an **H** field that propagate out from a point source (a source that is very small). The wave propagates equally in all directions, thus creating a spherical *wavefront*. On this wavefront, **E** and **H** are perpendicular to each other, and both are tangential to the surface of the wavefront. At any instant of time, the magnitude of **E** is the same everywhere on the wavefront. Also, the magnitude of **H** is the same everywhere on the wavefront. The direction of propagation of the wave is perpendicular to the spherical wavefront and in a direction radially out from the point source (along an arrow pointing outward). The vector **k** is often used to describe the direction of propagation of a wave. The relationship between **E**, **H**, and **k** is illustrated in Figure 3.1(a), which shows a cross section of some of the spherical wavefronts and **E**, **H**, and **k** at one point on a wavefront. **E**, **H**, and **k** are mutually perpendicular. (If the vector is pointing out of the paper, the tip of the vector is represented by a dot in a circle. If the vector is pointing into the paper, the tail of the vector is represented by a cross in a circle.) The vector **k** is in the direction a right-hand screw would move when **E** is turned into **H**. The right-hand rule described in Section 1.3 is often used to relate the directions of **E**, **H**, and **k**. The vector **k** is sometimes called the *ray* of the propagating wave, and the direction of that ray is often a convenient way of describing the path that the electromagnetic energy takes as it propagates through various regions. This is a particularly valuable visualization tool for the case when the wavelength is much smaller than the object, and will be discussed in more detail in Chapter 4.

The characteristics of the wave in the cross section of Figure 3.1(a) can be thought of as similar to the ripples in a pond of water when a pebble is dropped into it. The wave moves out with troughs and crests propagating away from the point at which the pebble entered the water, and the wave eventually dies out by dilution as it gets farther away from the

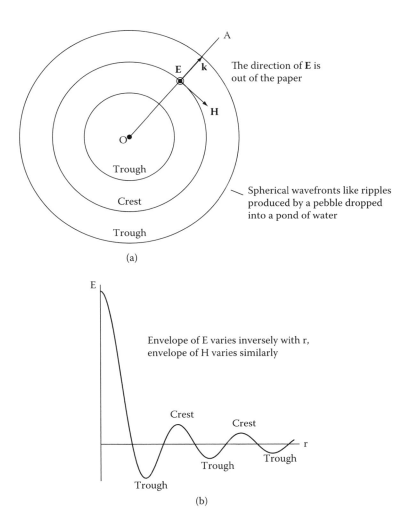

FIGURE 3.1
(a) Cross section of spherical wavefronts produced by a point source, showing **E** and **H** tangent to the wavefront, and the propagation vector **k** perpendicular to the wavefront. **E** is directed out of the paper and is therefore represented by a dot in a circle. **E**, **H**, and **k** are mutually perpendicular. (b) The variation of the magnitude of **E** as a function of radial distance r along the line OA at one instant of time. The peak magnitude of **E** varies inversely as r.

source. Similarly, Figure 3.1(b) shows that the peak magnitude of **E** varies inversely with the radial distance r away from the point source at one instant of time. The magnitude of **H** has a similar variation with r. Figure 3.2 shows how the vectors E and H vary at one instant of time with distance r along any radial line out from the point source. Figure 3.2 is like a snapshot of **E** and **H**. As time proceeds, the patterns shown in Figure 3.2 move out away from the point source; or, in other words, the wave propagates.

The ratio E/H is called the *wave impedance*. For spherical waves, the wave impedance is given by

$$\frac{E}{H} = Z = \sqrt{\frac{\mu}{\epsilon}} \text{ (ohms)} \tag{3.1}$$

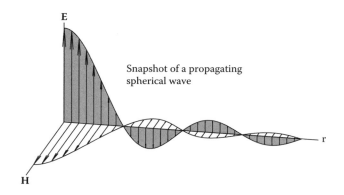

FIGURE 3.2
Pattern showing the **E** and **H** vectors of Figure 3.1 along a radial line at one instant of time. As time progresses, the pattern moves to the right.

where μ is the permeability and ε is the permittivity of the medium in which the wave is propagating. Since E and H are the complex values (magnitude and phase) of the electromagnetic wave, the wave impedance specifies the ratio of the **E** and **H** field magnitudes and gives information on their relative phases.

For a wave propagating in free space, where $\varepsilon = \varepsilon_0$ and $\mu = \mu_0$, the wave impedance is $Z = 377\Omega$ (Ω is the symbol for ohms), which is a real value. The fact that Z is strictly real in this case is important: its phase is zero, which means that there is zero phase shift between **E** and **H**, and therefore they are *in phase*. When **E** rises, so does **H**, and vice versa. This is true of all lossless materials. (For lossy materials, however, Z is complex, which means its phase is not zero and **E** and **H** are no longer in phase.)

For waves propagating in a material with permittivity ε, the impedance is lower than in free space. Since $\varepsilon = \varepsilon_0 \varepsilon_r$, where ε_r is the relative permittivity (see Section 1.6), then for nonmagnetic materials (in which $\mu = \mu_0$), the impedance can be written as

$$Z = 377/\sqrt{\varepsilon_r} \text{ (ohms)} \tag{3.2}$$

This means that **E** will be smaller for a given value of **H** than in free space. Biological materials are nonmagnetic ($\mu = \mu_0$), but have relatively high permittivities at these frequencies (see Appendix A) and are lossy.

The velocity of propagation of the wavefronts, called the phase velocity (see Section 1.11), is given by

$$v_p = 1/\sqrt{\mu\varepsilon} \text{ (m/s)} \tag{3.3}$$

For a wave propagating in free space, the phase velocity is usually designated by the letter c, and numerically,

$$c = 1/\sqrt{\mu_0 \varepsilon_0} = 3 \times 10^8 \text{m/s} \text{ (the speed of light in a vacuum),}$$

where μ_0 and ε_0 are the permeability and permittivity, respectively, of free space. As indicated by Equation 3.3, the phase velocity for a wave propagating in a material is lower than c because the permittivity and permeability of any material are greater than those of free space. Thus, materials are said to slow down propagating waves. Equation 3.3 can be written for nonmagnetic lossless materials (where $\mu = \mu_0$ and $\varepsilon'' = 0$) in the form

$$v_p = c / \sqrt{\varepsilon_r} \ \ (m/s) \tag{3.4}$$

For example, if a wave is propagating in a material with a relative permittivity $\varepsilon_r = 4$, its phase velocity will be half that of a wave propagating in free space.

As mentioned in the introduction to this section, spherical waves do not strictly exist in practice; there is no true physical point source of EM fields. The concept of spherical waves is often used to great advantage, however, because it is far simpler both mathematically and conceptually than most physical waves. It is therefore used to make calculations for idealized configurations to get approximate results for actual physical configurations. It is also used as a guide in constructing experiments and in interpreting experimental results.

Another major application where spherical waves come in handy is to approximate realistic extended sources as a sum of many very small sources. The field from the source is then calculated as a sum of the spherical fields associated with the many small sources (Huygen's principle). Several commercial software packages for calculating electromagnetic fields are based in some way on this approach.

For a spherical wave in a lossless material, the total power in each spherical wavefront is the same. But since the sphere increases in area proportional to r^2 as the wave propagates, the power density (power divided by area of the wavefront) decreases as a function of r^2 and the fields decrease as a function of r, where r is the distance from the source. Although spherical waves are not perfectly created in realistic applications, there are many applications and cases where this type of reduction in field strength is approximately observed.

At points far distant from the source in Figure 3.1, the radius of the wavefront is very large, making the curvature of the spherical wavefront so slight that the wavefront is approximately planar in a limited region. This is like the surface of the earth appearing approximately flat to us because the radius of the earth is so large compared to the limited region of our view. When the wavefronts become approximately planar, the wave approximates a *planewave*. Thus, we often assume the fields far from almost any source are planewave. This leads us into a discussion of planewaves, the subject of the next subsection.

3.2.2 Planewaves

As indicated by the name, planewaves are waves in which the wavefronts are planes. Figure 3.3 illustrates the **E**, **H**, and **k** for a planewave. As in spherical waves, **E**, **H**, and **k** for planewaves are mutually perpendicular, and the direction of **k** is the direction a right-hand screw would move when **E** is turned into **H**. **E** and **H** are tangential to the wavefronts, and **k** is perpendicular to the wavefronts. The magnitude and phase of **E** is the same everywhere on a given planar wavefront. Also, the magnitude of **H** is the same everywhere on a given planar wavefront. Figure 3.4 shows the **E** and **H** patterns at a given instant of time as a function of distance along any line perpendicular to the wavefronts. This figure is like a snapshot of **E** and **H** at a certain instant of time. As time proceeds, the pattern moves to the right. As indicated by Figures 3.3 and 3.4, the magnitudes of **E** and **H** do not decrease with distance in a truly lossless medium. This is unrealistic, but true of the idealized planewave.

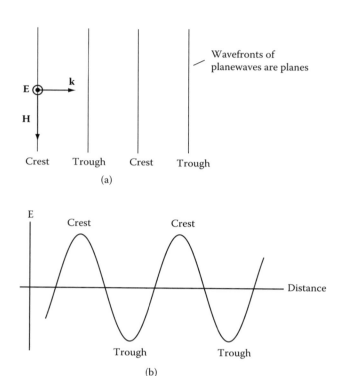

FIGURE 3.3
(a) Wavefronts of a planewave. The wavefronts are planes, only the edges of which are shown. **E** and **H** lie within a wavefront. The propagation vector **k** is perpendicular to the wavefronts and points in the direction of propagation. **E**, **H**, and **k** are mutually perpendicular. (b) The variation of the magnitude of **E** at one instant of time as a function of distance perpendicular to the wavefronts.

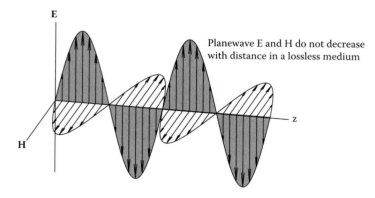

FIGURE 3.4
Pattern showing the **E** and **H** vectors of Figure 3.3 along a line in the direction of **k** at one instant of time. As time progresses, the pattern moves to the right.

What kind of source could produce planewaves? Because the **E** and **H** are constant everywhere on a planar wavefront, one would expect that only a source of infinite extent could produce planewaves. Such a source, of course, is unphysical. Perhaps the best approximations to planewaves are waves produced by finite sources when viewed in regions far from the source. Far enough away from almost any source the fields look like planewaves.

A planewave is similar to a spherical wave in many respects, as you might expect, because a spherical wave approximates a planewave in regions far away from the point source that produces the spherical wave. The wave impedance for a planewave is the same as the wave impedance for a spherical wave, as given by Equation 3.1. The phase velocity for a planewave is the same as the phase velocity for a spherical wave, as given by Equations 3.3 and 3.4. A significant difference between planewaves and spherical waves is that the peak magnitudes of the fields in spherical waves in lossless media vary inversely as distance, while the peak magnitudes of the fields in planewaves remain constant with distance.

If spherical waves and planewaves are unphysical, as we have repeatedly emphasized, why are they used so frequently? Because they are mathematically simple and can be used to understand the basic characteristics of wave interactions, as described in much of the remainder of this chapter. Simple spherical waves are often summed to produce more realistic waves, and planewaves represent fields far from most sources.

3.3 Wave Reflection and Refraction

Reflection and *refraction* are two important characteristics of wave behavior. When a propagating wave impinges on an interface between two different materials, or on an object, the wave can be partially reflected, refracted, or both. We will illustrate these behaviors with examples of planewaves impinging on planar interfaces, keeping in mind that planewaves do not actually exist physically, as explained in the previous section, but that they are extremely useful concepts for explaining important wave characteristics.

3.3.1 Planewave Reflection at Metallic Interfaces

Figure 3.5 illustrates reflection of a planewave that impinges on the interface of a perfectly conducting metallic halfspace (see Section 1.18 for a definition of halfspace). The incident planewave is represented by $\mathbf{E}_i$, $\mathbf{H}_i$, and $\mathbf{k}_i$, which are the electric field, magnetic field, and propagation vector, respectively, of a planewave propagating in free space toward the right. The wave is said to be normally incident on the conducting interface because the direction of propagation $\mathbf{k}_i$ is normal (perpendicular) to the metallic interface, which makes the $\mathbf{E}_i$ and $\mathbf{H}_i$ vectors tangential to this surface. Metal objects with perfect conductivity do not allow the wave to penetrate at all. They create a reflected wave that propagates in the opposite direction (in this case, to the left). The electric field vector of the reflected wave is inverted from the original at the interface; we say the reflected electric field is out of phase from the incident electric field. This means that the reflected field is 180 degrees out of phase from the incident field, and that the reflected electric field is the negative of the incident electric field at the metallic interface.

The fact that the field reflects off a metal surface is observed and used in a wide variety of applications. One obvious example that you have seen (literally) is light reflecting from a metallized mirror. Light is an electromagnetic wave, and you have undoubtedly seen light reflections from numerous metallic structures.

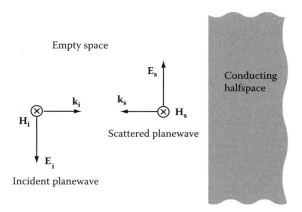

FIGURE 3.5

A planewave normally incident on a planar metallic interface and the reflected planewave produced by the metal. The subscript i stands for the incident wave, and the subscript s stands for the scattered (reflected) wave. The **H** fields are directed into the paper and are therefore represented by a cross in a circle.

Another application where reflections are used is in antenna design. Antennas require both a positive and a negative "arm," but many antennas (for example, your cell phone antenna) have only one arm. If a metal plate is placed under one arm of an antenna, the reflection from the plate makes it look to the antenna as if its other arm is "virtually" there underneath this ground plane, just as a reflection would appear to be behind the surface of the mirror. Radio stations often create a very large ground plane for their transmitting antennas by burying wires in a radial pattern like the spokes of a wheel at the base of the antennas. The spokes are close enough that they act as if the plane they create is solid (just as the screen of a microwave oven door reflects electromagnetic fields, as described in Section 1.17). Cell phones and other handheld devices are not large enough to have large, perfect ground planes underneath them, but the tops of their cases are generally metalized (typically plastic coated inside with metal paint). This creates a small ground plane that creates a partial reflection, somewhat like a small handheld mirror that reflects only part of the face.

Shielded metal rooms are used to test electromagnetic devices because they reflect external fields and do not allow them to interfere with the sensitive tests occurring inside. If the fields being tested are high, they also keep people on the outside of the chamber from being exposed to high electromagnetic fields. Power line workers who repair live power lines wear clothing with metallic covering to reflect the external field and keep them safe.

The reason for this reflection can be explained in terms of the simple boundary conditions originally explained in Section 1.12. When the planewave impinges on the metal and begins to propagate into it, the perfect conductivity forces the **E** field to go to zero. The **E** inside the metal must be zero because the conduction current density is given by $J_c = \sigma_c E$ (σ_c is conductivity), and since σ_c is infinite, **E** must be zero or else the conduction current density would be infinite, which is not possible. Because the boundary condition (Equation 1.18) requires the tangential **E** field to be continuous at the metallic interface, and because the **E** field in the metal is zero, the tangential **E** field at the metallic interface must also be zero.

This requirement causes a reflected wave to be generated, with the **E** field of the reflected wave equal and opposite to that of the incident wave at the interface. The actual total field at the boundary (and everywhere outside the metal) is the sum of the incident and reflected waves. The reflected wave, or scattered wave as it is often called, is represented in Figure 3.5 by $\mathbf{E}_s$, $\mathbf{H}_s$, and $\mathbf{k}_s$, which are the electric field, magnetic field, and propagation vector, respectively, of the scattered wave. At the boundary, the two equal and opposite electric fields add to zero, thus satisfying the boundary condition. Note, however, that the incident and scattered magnetic fields are both equal and in the same direction. Thus, the total tangential magnetic field is twice the incident magnetic field on the boundary.

The time-domain behavior of the planewave in front of a metal surface is particularly interesting. The total EM fields in the space to the left of the conducting halfspace in Figure 3.5 are the sums of the incident and scattered fields. Figure 3.6 shows the incident and reflected **E** field waves and their sum at nine different instants of time. The sum is zero at the metallic interface for all instants of time (because tangential electric fields are always zero on perfectly conducting metal). If you look closely at Figure 3.6, you will see that at certain positions in front of the metallic interface, the sum of the incident and reflected waves is zero for all nine instants of time. This is shown more clearly in Figure 3.7(a), which shows just the sum of the incident and reflected wave electric fields as a function of distance z at all nine instants of time superposed on the same graph. At points that lie at half-wavelength intervals in front of the metallic surface, the total **E** is zero at all nine instants of time. It turns out that at these points the **E** is zero at all instants of time, not just the nine shown in the figure. These zero values of **E** are called *nulls*. The nulls occur

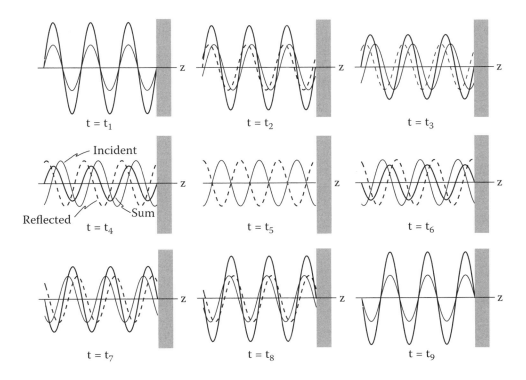

FIGURE 3.6
The electric fields of the incident wave (light solid lines) of Figure 3.5, the scattered (reflected) wave (dashed lines), and the sum of the incident and scattered waves (bold lines) as a function of distance, at nine different instants of time. At t_1 and t_9, the scattered wave lies on top of the incident wave. At t_5 the sum of the waves is zero. The gray rectangle represents the position of the planar metallic interface.

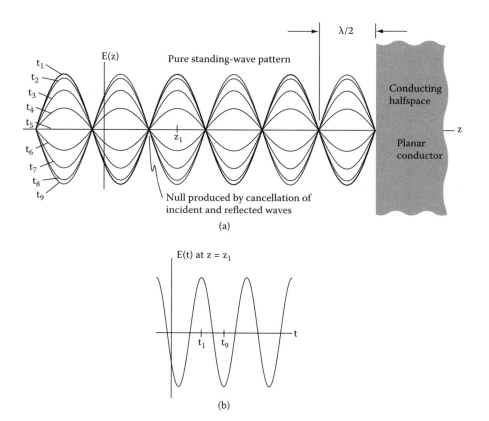

FIGURE 3.7
(a) The sums of the electric fields of the incident and scattered waves of Figure 3.6 for nine instants of time all plotted on the same graph as a function of distance in front of the planar metallic object. This is an illustration of a standing wave. The E at t_1 is shown in bold to illustrate the typical pattern. (b) The electric field at position z_1 as a function of time.

because the incident and scattered **E** fields are equal in magnitude and opposite in sign at those points, so that they cancel each other out at all times. This cancellation occurs because the incident and scattered waves propagate with the same phase velocity, but in opposite directions, and because the incident and scattered **E** must add to zero at the metallic interface.

If you imagine the electric field patterns at different times in Figure 3.7(a) appearing sequentially as a "movie," you can see why the wave in front of a metal surface is called a *standing wave*. The total electric field (sum of the incident and reflected waves) moves up and down at any given point (except the nulls, which remain at zero) in a sinusoidal fashion. If you measured the electric field at any individual point in front of the planewave, you would find that its amplitude is a sinusoidal function of time. Figure 3.7(b) shows the sinusoidal time variation of the amplitude at one point, $z = z_1$.

Figure 3.8 shows a similar standing wave pattern for the total **H** fields. Because of the mutually orthogonal relations between **E**, **H**, and **k** in each of the waves, the incident and scattered **H** fields *add* at the metallic interface instead of cancelling, and the first null in **H** occurs a quarter-wavelength back from the metallic interface, and then at half-wavelength intervals thereafter.

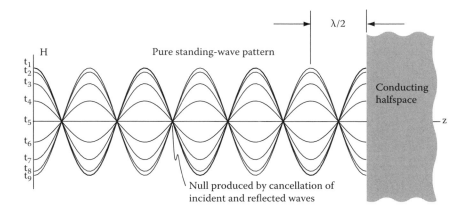

FIGURE 3.8
The sums of the magnetic fields of the incident and scattered waves of Figure 3.6 for nine instants of time all plotted on the same graph as a function of distance in front of the planar metallic object. This is an illustration of a standing wave. The H at t_1 is shown in bold to illustrate the typical pattern.

Standing wave patterns are often represented by just the *envelope* of the pattern, as shown in Figure 3.9. The envelopes of the **E** and **H** standing waves clearly show the positions of the nulls and the maximum values that the **E** and **H** attain.

When a planewave is incident on a metal object at an angle other than normal (or head on, where $\theta_i = 0°$), even more interesting field patterns occur. This is referred to as being obliquely incident and is shown in Figure 3.10. The angle θ_i that the propagation vector ($\mathbf{k}_i$) makes with a normal to the metallic surface is called the *angle of incidence*. As with normal incidence, the boundary conditions at the metallic interface require the tangential electric field to be zero there, again producing a scattered wave. The field reflects off the metal surface at an angle θ_s that is equal to the angle of incidence θ_i. This means that the propagation vector of the scattered wave $\mathbf{k}_s$ makes an angle θ_s with the normal to the surface that is equal to the angle of incidence. Once again, the sum of the electric fields of the incident and scattered waves is a standing wave, but for oblique incidence the nulls are farther apart than one half wavelength. The greater the angle of incidence, the farther apart the nulls.

Figure 3.11 shows the incident and scattered waves at one instant of time in terms of the peaks and troughs—the peaks as black lines and the troughs as gray lines. At points where black and gray lines intersect, the sum of the **E** in the two waves is zero. Although intermediate values of **E** between the peaks and troughs are not explicitly shown in Figure 3.11, the sum of the incident and scattered **E** is zero everywhere along the dashed lines at every instant of time. Thus, a standing wave pattern occurs for oblique incidence just as it does for normal incidence, but the nulls in the standing wave for oblique incidence are farther apart than for normal incidence, as mentioned above.

Another way of illustrating the electric fields of the incident and reflected waves and their sum is shown in Figures 3.12 to 3.14. In these figures, the amplitude of the **E** is represented by shades of gray, with white being maximum and black being minimum. Figure 3.12 shows the **E** of the incident wave at one instant of time, Figure 3.13 the **E** of the scattered wave, and Figure 3.14 the sum of the two. As time progresses, the patterns move according to the propagation directions of the waves shown by the two arrows in Figure 3.11, but the nulls in the total **E** always occur at positions along the dashed lines in Figures 3.11 and 3.14.

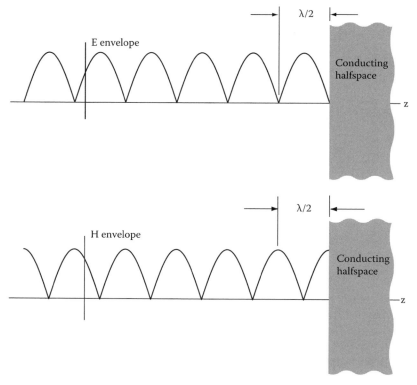

FIGURE 3.9
Envelopes of the **E** and **H** standing waves of Figures 3.7 and 3.8.

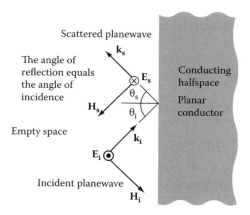

FIGURE 3.10
A planewave obliquely incident on a perfectly conducting planar metallic interface and the reflected (scattered) planewave produced by the metal. The subscript i stands for the incident wave, and the subscript s stands for the scattered (reflected) wave. The angle of reflection θ_s is equal to θ_i, the angle of incidence.

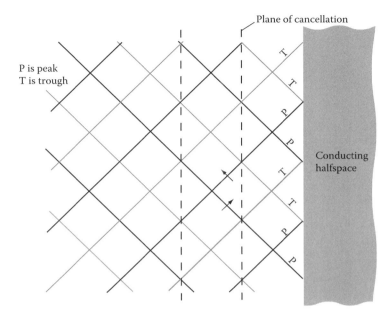

FIGURE 3.11
Wavefronts of the incident and scattered planewaves of Figure 3.10 at one instant of time. P stands for peak, and T stands for trough. The dashed lines show where the peaks and troughs add to zero. The respective wavefronts propagate in the directions of the arrows as time progresses.

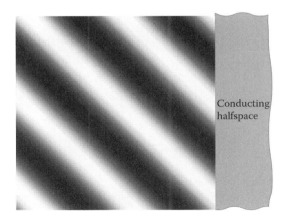

FIGURE 3.12
The amplitude of the **E** of the incident wave of Figure 3.10 at one instant of time. The peaks are white, and the troughs are black. Values in between peaks and troughs are shown in various shades of gray.

FIGURE 3.13
The amplitude of the **E** of the scattered wave of Figure 3.10 at one instant of time. The peaks are white, and the troughs are black. Values in between peaks and troughs are shown in various shades of gray.

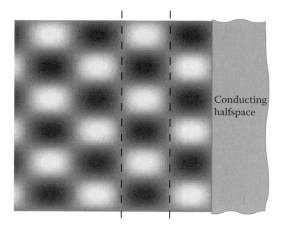

FIGURE 3.14
The amplitude of the sum of the **E** of the incident wave and the **E** of the scattered wave of Figure 3.10 at one instant of time. The maxima are white, and the minima are black. Values in between the maxima and minima are shown in various shades of gray. The dashed lines indicate planes of cancellation of the incident and scattered waves.

GROUNDING AND SHIELDING IN ELECTROMAGNETICS

Grounding is an important concept in electrical engineering. Voltages are measured relative to a ground, which is usually treated as zero voltage. Currents flow from a positive voltage down to a ground. Properly grounding equipment is important to prevent shocks and hazard.

At low frequencies (Chapter 2), one electrode has a positive voltage and the other is treated as the ground. Most often the ground electrode is attached on the body at some distance from the first electrode. The current will pass from the positive electrode to the ground electrode. At low frequencies, this current flows along the path of least resistance. It will tend to flow and concentrate in the high-water-content, high-conductivity tissues. At low frequencies, a ground can be created in many different ways; any metallic contact will create a ground. A rod, wire, or clip can all be used to create a sufficient ground for low frequencies.

At middle and higher frequencies, the situation is quite different. The current follows the path of least inductance, not least resistance. A single wire often will not be sufficient to provide a ground, and the current will choose *(continued on next page)* many other paths to reach the ground. The current is no longer constrained to just the metallic contact points. Ground plates are commonly used, and multiple parallel connections are attached to them. Sometimes this creates another problem—ground loops. The current can pass up some of the parallel connections and down others, creating loops that cause delays in parts of the signal and make it spread out over time. Pulsed electromagnetic fields have both high- and low-frequency components, and they should be treated as high-frequency fields when considering grounding.

Shielding is another important aspect of high-frequency devices. Since high-frequency fields are no longer contained in the metal and can be transmitted through space, shielding is important to provide protection from electromagnetic interference (EMI). Since electric fields cannot penetrate metal, putting the device to be shielded in a completely enclosed metal container or can (with the seams electrically sealed) will prevent the device from receiving any signals from outside of the can. Of course, we usually want to get a signal in or out of the device in the can, but as soon as we make a hole in the can to run a wire, the can is now "leaky" with respect to electromagnetic fields. As with the microwave oven door described in Section 1.17, the size of the hole determines which frequencies will pass through and which will be rejected. So typically we use a shielded cable such as a coax to reach the device inside the can. The outer shield of the cable is connected (with no holes) to the body of the can, with the inner conductor of the coax reaching the device inside the can. We also have to be careful to do the same on the other end of the cable.

Full shielding is often important for very sensitive electromagnetic measurements done in a laboratory. In this case, the shielding is provided by a metal mesh or plates completely surrounding the entire room: a shielded room. The measurement equipment and power supply are often left outside the room, and the receiving antenna and device being measured are inside the room. This protects the measurements from corrupting RF signals such as radio stations and cell phone signals, and makes it possible to receive very tiny signals without excessive outside noise.

3.3.2 Planewave Reflection and Refraction at Dielectric Interfaces

As explained in the previous section, infinite conductivity causes the **E** inside a perfect conductor to be zero, and since the boundary conditions require the tangential **E** to be continuous at interfaces, the tangential **E** must be zero at the surface of a perfect conductor. This boundary condition causes a reflected, or scattered, wave to be produced when a wave is incident on a perfect conductor. A reflected wave is also produced when a wave is incident on a good, but not perfect, conductor because the **E** field will be very small, but not zero, in the good conductor. Quite a different effect occurs in a dielectric, in which the internal **E** field is not zero. The boundary conditions (Equations 1.17 and 1.18) at the surface of a dielectric also produce a scattered wave, but in addition, a wave is transmitted into the dielectric.

Figure 3.15 shows a planewave normally incident on a dielectric halfspace, along with the resulting scattered and transmitted waves. The boundary conditions require that the sum of the incident and scattered **E** be equal to the transmitted **E** at the dielectric interface. As shown in Figure 3.16(a) for several instances of time during an oscillatory cycle, the sum of the incident and scattered electric fields in front of the dielectric produces a

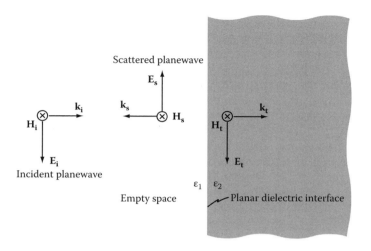

FIGURE 3.15

A planewave normally incident on a planar dielectric interface and the scattered and transmitted planewaves. The subscript i stands for the incident wave, the subscript s stands for the scattered (reflected) wave, and the subscript t stands for the wave transmitted into the dielectric.

pattern that is similar to the standing wave in front of the metal (Figure 3.7). However, in the pattern produced by the dielectric interface, there are no nulls where the total wave is completely zero at all times, but there are minima at regular intervals. Since the envelope never goes to zero, this pattern is often referred to as a *partial standing wave*.* The **E** transmitted into the dielectric is a traveling wave, and its envelope is uniform everywhere inside the dielectric.

The envelopes of the waves in Figure 3.16(a) are shown in Figure 3.16(b). Because the amplitude of the reflected wave increases and the minima become smaller as the permittivity of the dielectric increases, the wave pattern approaches that of a pure standing wave for dielectrics of very high permittivity. Also, as the permittivity of the dielectric increases, the amplitude of the transmitted wave decreases. Thus, a high-permittivity dielectric reflects waves similarly to a good conductor.

When a planewave is obliquely incident on a dielectric halfspace (Figure 3.17), the angle of reflection is equal to the angle of incidence, as it is with a conductor (Figure 3.10). The angle of refraction (or transmission) θ_t depends on θ_i and the permittivities of the incident medium ε_1 and of the dielectric ε_2. For a given θ_i and ε_1, θ_t decreases as the permittivity ε_2 increases. The relation between these three quantities is a famous one known as *Snell's law of refraction*:

$$\sqrt{\varepsilon_1}\,\sin\theta_i = \sqrt{\varepsilon_2}\,\sin\theta_t \tag{3.5}$$

where ε_1 is the permittivity of the medium in which the incident and scattered waves are found, and ε_2 is the permittivity of the dielectric into which the transmitted wave propagates.

When ε_2 is less than ε_1—that is, when the incident wave impinges on a medium of lesser permittivity—a special effect called *total internal reflection* can occur. This corresponds to θ_t

---

* A partial standing wave is a superposition of a pure standing wave and a traveling wave; it has an envelope that is a combination of that seen in Figure 3.9 plus a straight-line envelope of a traveling wave.

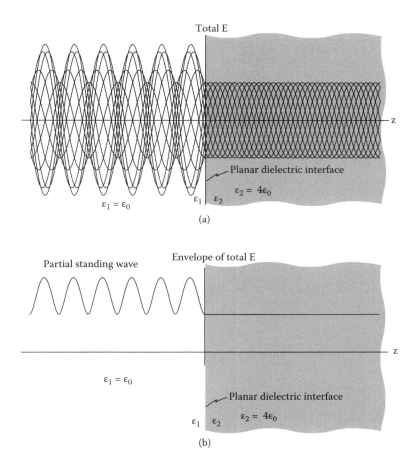

FIGURE 3.16
(a) The total (sum of incident and scattered) **E** to the left of the planar dielectric interface and the transmitted **E** in the dielectric media on the right at thirteen different instants of time for the waves of Figure 3.15, all plotted on the same graph as a function of distance. This is an example of a partial standing wave. $\varepsilon_1 = \varepsilon_0$ and $\varepsilon_2 = 4\,\varepsilon_0$. (b) Envelope of the **E** of (a).

being equal to or greater than 90°, for which angles the wave would not be transmitted at all into the second medium. The angle of incidence for which $\theta_t = 90°$ is called the *critical angle*, θ_{ic}. From Equation 3.5 we can solve for θ_{ic} using the fact that when $\theta_t = 90°$, $\sin \theta_t = 1$. Substituting this value in Equation 3.5 and solving for $\sin \theta_{ic}$ gives

$$\sin \theta_{ic} = \sqrt{\varepsilon_2/\varepsilon_1} \tag{3.6}$$

For example, if medium 2 is air and medium 1 has a relative permittivity of 4, then $\varepsilon_2/\varepsilon_1 = 1/4$, and $\theta_{ic} = 30°$. Then, as illustrated in Figure 3.18, only waves within a cone of 60° would be transmitted out from medium 1 (with reduced amplitude), and all others would be totally internally reflected.

A number of important characteristics of wave reflection and refraction by multiple interfaces are illustrated by the simple case shown in Figure 3.19, a planewave propagating to the right and normally incident on a dielectric slab. By dielectric slab, we mean an object

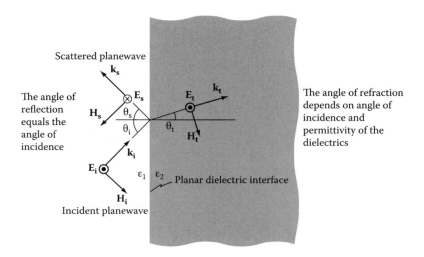

FIGURE 3.17
A planewave obliquely incident on a planar dielectric interface and the scattered and transmitted planewaves. The subscript i stands for the incident wave, the subscript s stands for the scattered (reflected) wave, and the subscript t stands for the wave transmitted into the dielectric. Snell's law of refraction describes the relationship between the angles.

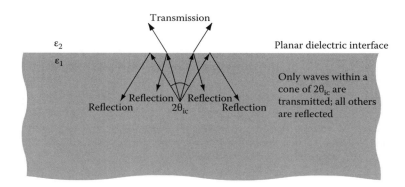

FIGURE 3.18
Waves incident on a lower-permittivity medium from a higher-permittivity medium at two angles, one that is less than the critical angle, where there is some transmitted portion, and one at the critical angle, where there is no longer any wave transmitted into the upper region. If medium 2 is air and if medium 1 has a relative permittivity of 4, the critical angle is 30°. For angles of incidence greater than 30°, the waves will not be transmitted into the air, but will be totally reflected back into medium 1.

that is of a specified thickness in the direction of the $\mathbf{k}_{1a}$ vector in Figure 3.19, and infinite in extent in all other directions. When the wave impinges on the left interface of the slab, part of it is transmitted into the slab and part of it is reflected. The part that is transmitted impinges on the right interface, where part of it is transmitted (continuing through the interface to the right) and part is reflected (to the left). The part that is reflected travels to the left and impinges on the left interface, where part of it is transmitted and part of

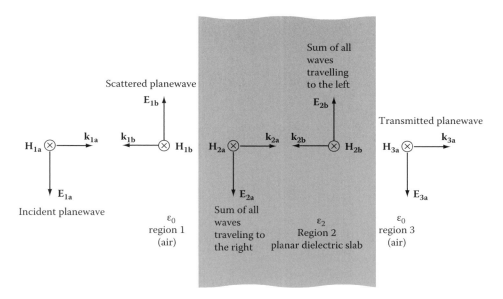

FIGURE 3.19

A planewave propagating to the right ($\mathbf{E}_{1a}$, $\mathbf{H}_{1a}$, $\mathbf{k}_{1a}$) is normally incident on a planar lossless (zero effective conductivity) dielectric slab. The left interface of the slab partially transmits and reflects the planewave. The wave transmitted into the slab is partially transmitted and reflected at the right interface. The wave in the slab that is reflected at the right interface travels back to the left interface and is there partially transmitted and reflected. The process continues until the steady state is reached, which consists of multiple partial transmissions and reflections at each interface. The subscripts a on the $\mathbf{E}$, $\mathbf{H}$, and $\mathbf{k}$ in each region represent the total of all the waves traveling to the right. The subscripts b represent the total of all the waves traveling to the left.

it is reflected again. This continues until the steady state is reached. Each transmission and reflection is progressively smaller. Steady state is when the additional reflections and transmissions are so small we cannot measure them or they are below some minimum value of interest to us. The steady state then consists of the result of multiple transmissions and reflections at each interface. In the figure, the subscript a represents the sum of all the waves traveling to the right, and the subscript b all those traveling to the left. The combination of all these waves results in a wave pattern similar to those shown in Figure 3.16.

Figure 3.20 shows the electric field envelopes of these wave patterns as a function of the thickness of a lossless (zero effective conductivity) dielectric slab having a relative permittivity of 4. The thickness is given in terms of the wavelength inside the dielectric slab, λ_d. Because the dielectric decreases the velocity of propagation of the planewave according to Equation 3.4, the wavelength in the dielectric, as obtained from Equation 1.15, is given by

$$\lambda_d = \frac{c}{f\sqrt{\varepsilon_r}} = \frac{\lambda}{\sqrt{\varepsilon_r}} \tag{3.7}$$

where λ is the wavelength in free space ($\lambda = c/f$) and ε_r is the relative permittivity of the dielectric. The dielectric thus slows the wave and decreases the wavelength.

As indicated by the graphs in Figure 3.20, the thickness of the slab has a drastic effect on the wave patterns. When the slab is a quarter-wavelength thick, the incident wave is

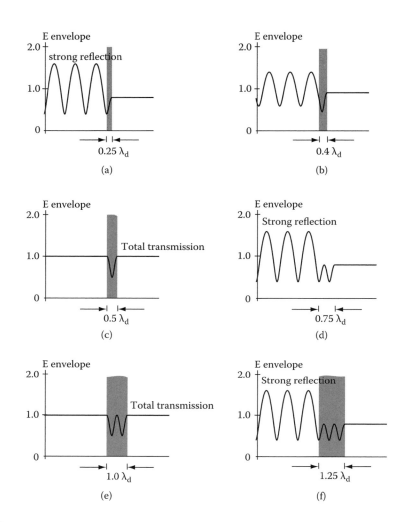

FIGURE 3.20
Electric field envelopes for the configuration of Figure 3.19 for six different widths of the dielectric slab. The widths are expressed in terms of the wavelength inside the dielectric slab (λ_d). The relative permittivity of each dielectric slab is 4, and the effective conductivity is zero (it is a lossless dielectric).

strongly reflected, as indicated by the envelope in Figure 3.20(a) in the free-space region to the left of the slab. The straight line to the right of each slab represents the envelope of the transmitted wave. The height of this envelope indicates how much of the incident wave (which has an envelope of unity height) is transmitted through the slab. As shown in Figure 3.20(a), the transmitted height is less than unity. When the thickness is increased to 0.4 λ_d, more of the incident wave is transmitted through the slab, as shown in Figure 3.20(b). When the thickness is increased to one-half a wavelength, a striking effect occurs. The entire incident wave is transmitted through the slab, as shown in Figure 3.20(c). Reflections still occur inside the slab, but no reflections occur at the left interface, as indicated by the flat line to the left of the slab, and the envelope of the transmitted wave to the right of the slab is the same height as that of the incident wave.

As the thickness is increased to 0.75 λ_d, strong reflection again occurs. Then for a thickness of one wavelength, the entire incident wave is again transmitted through the slab.

Finally, at 1.25 λ_d, the same reflection occurs to the left as occurs for 0.25 λ_d, but multiple maxima and minima occur inside the slab. For multiples of a half wavelength, all of the wave is transmitted through the slab. For odd multiples of a quarter wavelength, strong reflection and low transmission occur.

The characteristics illustrated by Figure 3.20(c) and (e), in which zero reflection and maximum transmission occur, have important practical applications because the same kind of characteristics occur for real waves and nonplanar dielectric objects. For example, dielectric covers (radomes) for radar antennas are made multiples of a half wavelength in thickness to transmit the radar signals through with minimum reflection.

MEASURING DIELECTRIC PROPERTIES

In this frequency range, the electrical properties of materials are most commonly measured with a coaxial probe. A simple coaxial line is cut off at the end and left "open," then pressed against or immersed in the material being measured. The outer conductor (shield) of the coaxial probe is typically connected to a larger guard electrode. The electric field lines of an applied probing signal (which are always perpendicular to metal) are perpendicular to both the end of the center conductor and the guard electrode. Thus, they bend around the open end of the coaxial line as shown in Figure 3.21. These electric field lines interact with the material being measured and produce a reflected field that returns down the coaxial transmission line. The characteristics of this reflected field (magnitude and phase) are a function of the material's permittivity and conductivity. The probing signal is typically swept in frequency to determine the electrical properties as a function of frequency. Most commonly, this dielectric measurement probe is connected to a network analyzer that both provides the probing signal and contains software to determine the material's permittivity and conductivity.

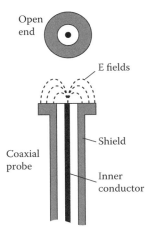

FIGURE 3.21
The open end of a coaxial cable probe used for measuring the conductivity and permittivity of materials.

3.4 Waves in Lossy Media

Lossy media are those in which the conductivity, or effective conductivity (see Section 1.14), is not negligible. If the conductivity is high, as it is in metals, we call the material a *conductor*. If the conductivity is relatively low, we call the material a *lossy dielectric*. In this section we first discuss waves in metals, and then we discuss waves in lossy dielectrics.

3.4.1 Waves in Metals

As explained in Section 3.3.1, the **E** inside a *perfect* conductor (infinite conductivity) must be zero because the conduction current density is given by $\sigma_c \mathbf{E}$, and since σ_c is infinite, **E** must be zero. Otherwise, the conduction current density would be infinite, which is not possible. If the conductivity is high, but not infinite, the **E** inside the metal is not forced to be identically zero, but it is small. As a wave propagates in a good, but not perfect, conductor, the conductivity causes the wave to rapidly attenuate because the **E** fields of the wave transfer energy to the charges in the conductor.

Figure 3.22 shows the **E** field envelope of a planewave impinging normally on a conducting halfspace. As indicated by the envelope, the interface creates a reflected wave that causes a partial standing wave to the left of the interface. The conductivity of the halfspace also attenuates the wave as it propagates into the halfspace. We set the conductivity in Figure 3.22 to be relatively low so that the nature of the attenuation is more evident.

When the conductivity is very high, the attenuation is so rapid that the **E** and **H** fields become essentially zero within a very small distance of the interface. The associated currents that flow in the material are therefore confined to a very thin layer near the surface. This is called the *skin effect*. The skin depth δ is defined as the depth at which the **E** and **H** have attenuated to 1/e (0.37) of their values at the surface (e = 2.718 is the base of the natural logarithm). The skin depth is shown on the diagram in Figure 3.22.

For a planewave impinging on a conducting halfspace, the skin depth is given by

$$\delta = \sqrt{2/\omega\mu\sigma_c} \ \ \text{(m)} \tag{3.8}$$

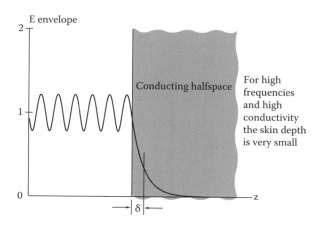

FIGURE 3.22
E field envelope for a planewave normally incident on a conducting halfspace. δ is the skin depth.

where ω is the radian frequency, μ is the permeability, and σ_c is the conductivity. When ω and σ_c are both high, the skin depth is very small. For example, the conductivity of copper is $\sigma_c = 5.80 \times 10^7$ S/m and the permeability is $\mu = 4\pi \times 10^{-7}$ H/m. Thus, in copper at a frequency of 10 GHz ($\omega = 2\pi \times 10^9$ rad/s), the skin depth is given by

$$\delta = \sqrt{2/(2\pi 10 \times 10^9 \times 4\pi 10^{-7} \times 5.80 \times 10^7)} = 0.66 \times 10^{-6} \text{ m}$$

The skin depth in other good conductors is similarly very small at high frequencies. At lower frequencies, the skin depth is correspondingly greater, as given by Equation 3.8. One of the important impacts of the skin effect is seen in coated conductors. If a copper wire is coated with tin in order to make it easier to solder, for instance, the currents can be contained in the tin rather than the copper at high frequencies due to the skin effect. Tin is much more lossy than copper, and therefore whatever system is created with this wire is not as efficient as it would have been with copper alone.

3.4.2 Waves in Lossy Dielectrics

The wave pattern for a planewave impinging on a lossy dielectric halfspace is similar to that shown in Figure 3.22 for a conducting halfspace. As mentioned above, we usually refer to conductors as those materials having relatively high conductivity and lossy dielectrics as those materials having relatively low conductivity. The definition for the skin depth is the same for both kinds of materials. Thus, the wave pattern in Figure 3.22 could be for a poor conductor or for a lossy dielectric. The skin depth as a function of frequency for tissue (a lossy dielectric) is shown in Figure 1.40, where the conductivity is the effective conductivity (see Section 1.14 and Equation 1.27) that varies with frequency.

Wave patterns produced by a lossy dielectric slab of varying thickness are shown in Figure 3.23. Comparison with those of Figure 3.20 for the lossless slab show that the loss has a significant effect on the wave patterns. The envelopes inside the slab show how the loss decreases the amplitudes of the waves as they travel through the slab. A particularly significant effect of the loss is that zero reflection no longer occurs when the thickness of the slab is a multiple of a half wavelength, as indicated by the patterns of Figure 3.23(c) and (e).

3.4.3 Energy Absorption in Lossy Media

The *Poynting vector* describes the power density stored in the **E** and **H** fields in a wave. For sinusoidal planewaves, the magnitude of the time-average Poynting vector is given by

$$P = \frac{E_{rms}^2}{\sqrt{2}} \sqrt{\frac{\varepsilon'}{\mu}} \sqrt{1 + \sqrt{1 + (\sigma_{eff}/\omega\varepsilon')^2}} \quad (W/m^2) \tag{3.9}$$

in terms of the **E** of the planewave, or equivalently by

$$P = \frac{H_{rms}^2}{\sqrt{2}} \sqrt{\frac{\mu}{\varepsilon'}} \frac{\sqrt{1 + \sqrt{1 + (\sigma_{eff}/\omega\varepsilon')^2}}}{\sqrt{1 + (\sigma_{eff}/\omega\varepsilon')^2}} \quad (W/m^2) \tag{3.10}$$

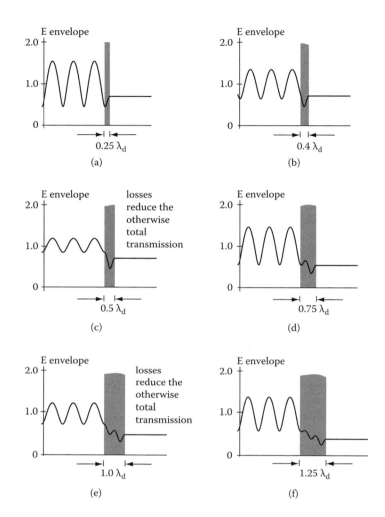

FIGURE 3.23
The same graphs as in Figure 3.20, but for this case the dielectric is lossy.

in terms of the **H** of the planewave. The Poynting vector may be thought of as describing the power per unit area that is traveling through, and therefore can be transferred to, the material in which the wave is propagating. These two relations are used in dosimetry (see Chapter 5) in relating the energy absorbed in models of humans and other animals to the fields and power of the incident waves.

The specific absorption rate (SAR) (see Section 1.16) that describes the time rate of energy transferred from a planewave to a material (per unit mass) at a given point is given by Equation 1.41. Because the envelope of E is the peak value of the **E** field at each point, and since the rms value is equal to 1/2 the peak value (see Section 1.10), the SAR is proportional to the square of the envelope. Thus, the envelopes inside the dielectric slabs of Figure 3.23 indicate that the SARs can vary significantly from point to point inside the slabs.

Figure 3.24 shows examples of the SAR inside lossy dielectric slabs. Figure 3.25 shows the same information in terms of a gray-scale plot on a larger scale. The gray-scale display dramatizes the strong variation of the SAR with position in the slab (maximum SAR is

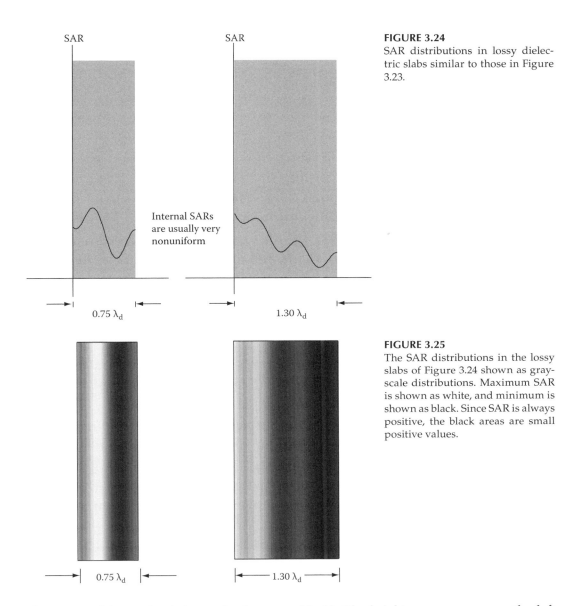

FIGURE 3.24
SAR distributions in lossy dielectric slabs similar to those in Figure 3.23.

FIGURE 3.25
The SAR distributions in the lossy slabs of Figure 3.24 shown as grayscale distributions. Maximum SAR is shown as white, and minimum is shown as black. Since SAR is always positive, the black areas are small positive values.

shown as white, and minimum is shown as black). The brighter areas are near the left sides of the slabs, where the EM wave is incident, and the darker areas are toward the right side of the slabs, where the EM wave has been attenuated by the losses in the slabs. The SAR pattern is a strong function of the thickness of the slab compared to the wavelength of the EM radiation.

The bright areas are often referred to as *hot spots*, although this is not precise nomenclature because *hot* refers to temperature, and temperature inside a body exposed to EM radiation depends not only on the SAR, but also on the thermal properties of the body. For example, in animals, the temperature is a strong function of the thermal regulatory mechanisms of the body, such as increased blood perfusion.

These examples illustrate the general characteristics of SAR distributions inside the bodies of humans and other animals exposed to EM radiation. The distribution inside the body is a strong function of the size of the body compared to the wavelength of the EM radiation, and when the body size is of the same order of magnitude as the wavelength, the SAR distribution is usually very nonuniform. The nonuniformity is caused by the multiple reflections and refractions at the interfaces between materials of different electrical properties. This resonance effect is discussed further in Section 3.6. While the examples given here are for homogeneous slabs, the nonuniformities are further increased by inhomogeneities in objects, like the inhomogeneous tissues of animal bodies.

3.5 Transmission Lines and Waveguides

In the sections above, we discussed free-space wave propagation, reflection, and refraction. In this section we describe wave propagation along guiding systems (transmission systems), such as two-wire transmission lines, coaxial cables, and hollow pipes called waveguides. Such systems are widely used to transmit EM signals when the frequency is high enough that the wavelength of the EM fields is of the same order of magnitude as the size of the transmission system.

When waves propagate along guiding systems, the **E** and **H** fields exist in characteristic combinations called *modes*. The three most common modes are transverse electromagnetic (TEM), transverse electric (TE), and transverse magnetic (TM). In the TEM mode, both **E** and **H** are transverse (perpendicular) to the direction of propagation. In the TE mode, **E** is transverse to the direction of propagation, but **H** is not. In the TM mode, **H** is transverse to the direction of propagation, but **E** is not. A planewave, for example, is a TEM wave because, as explained in Section 3.2.2, **E** and **H** are both perpendicular to the propagation vector **k**, which lies along the direction of propagation.

TEM modes can exist on structures consisting of two conductors, such as two wires or a coaxial cable (which consists of a wire centered inside a hollow conductor), but they cannot exist in hollow pipes as TE or TM modes can. In this section, we first discuss TEM transmission systems, which are usually referred to as transmission lines, and then we discuss waveguides, which propagate combinations of TE and TM modes.

3.5.1 TEM Systems

In Section 1.17, we stated that when the frequency is high enough that the wavelength is comparable to the size of the system, voltage could be defined only in very special cases. One of those special cases is TEM modes. As explained in Section 1.2, a unique voltage can be defined between two points when **E** is a conservative field. From Equation 1.3, it can be shown that when **E** and **H** lie in the same plane, as they do in TEM modes, **E** is conservative. Therefore, voltage can be uniquely defined for TEM modes.

TEM modes can exist on transmission systems consisting of two conductors. Several examples of such systems are shown in Figure 3.26. In each case a source, such as a voltage source or current source, produces a voltage difference between the two conductors and causes current to flow in the conductors. Other modes can also exist on these transmission systems, but when these systems are used, they are usually designed so that the TEM mode dominates, and the other modes are negligible.

Figure 3.27 shows the **E** field pattern at one instant of time between the conductors of the stripline of Figure 3.26(c). A unique potential difference exists between the two conductors

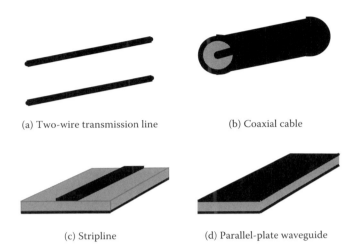

(a) Two-wire transmission line   (b) Coaxial cable

(c) Stripline   (d) Parallel-plate waveguide

FIGURE 3.26
Examples of two-conductor transmission systems. In each case the conductors are shown in black, and some kind of dielectric material (could be air) is placed between the conductors.

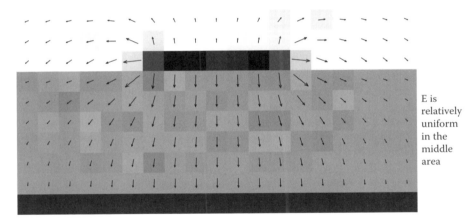

E is relatively uniform in the middle area

FIGURE 3.27
Two-dimensional calculations of the **E** field pattern between the two conductors of the stripline shown in Figure 3.26(c). A sinusoidal current source (not shown) is connected between the two conductors. The **E** fields are shown at an instant of time when the source current is zero. The **E** fields are negligibly small at an instant of time when the source current is maximum. This corresponds to a 90° phase shift between the **E** fields and the source current. For clarity, only the central region of the stripline is shown. The fields to the left and to the right of the region shown are very small.

because the same amount of work is required to move a charge from one conductor to the other along any path between the two (see Section 1.2). The same is true for all TEM systems. Since a unique potential difference can be defined for TEM transmission systems, transmission along them is usually described in terms of voltage $V(z)$ between the conductors and current $I(z)$ in the conductors (Figure 3.28), each described as a propagating wave. These propagating voltage and current waves on all TEM systems are described by the same set of equations, which are called the transmission-line equations. Thus, the characteristic behaviors that result from the solution of these transmission-line equations apply to all TEM systems, and we describe these next.

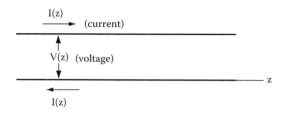

FIGURE 3.28
Voltage and current on a two-wire transmission line. V(z) is the potential difference of one wire with respect to the other at a point z on the transmission line. I(z) is the current in the wire at the point z, with equal and opposite currents in the two wires at that point on the line.

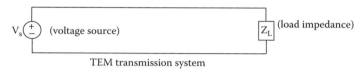

FIGURE 3.29
A diagram of a TEM transmission system with a source (voltage source shown here) connected to the left end, and a system or device (represented by a load impedance) connected to the right end. A current source is another typical source that could be connected to a transmission system.

A typical transmission-line configuration is a source applied at the left end of the transmission line, which causes voltage and current waves to propagate along the line, usually to carry information of some kind (a signal), or to transmit energy to some device or system (often called the load, which is represented as a load impedance Z_L). This configuration is diagrammed in Figure 3.29. The source is often a sinusoidal function of time, although it could be any function of time. When it is a sinusoidal function of time, the phasor transform (see Section 1.13) is usually used to solve for the voltages and currents.

The source generates a voltage wave and a current wave that propagate to the right. If the transmission line were infinitely long (a physical impossibility, but a useful concept), only waves traveling to the right would exist. When the line is finite, any kind of discontinuity in the line—such as a change in the size and shape of the conductors, a change in the dielectric properties between the conductors, or some device connected between the conductors—will cause reflections, which are waves traveling to the left. These reflections are like those described for planewaves (Section 3.3).

The ratio of the phasor voltage (the sum of the voltage waves traveling to the right and the ones traveling to the left) to the phasor current (the sum of the current waves traveling to the right and the ones traveling to the left) at any point on the line is called the *impedance*. In general, the impedance is a function of position along the transmission system; that is, it varies with position. When only waves traveling to the right exist (or to the left, but not both), such as in an infinitely long line, the ratio of the voltage to the current has the same value at any point on the transmission system, and this ratio is called the *characteristic impedance* of the transmission system. The characteristic impedance is usually designated by the symbol Z_0. The characteristic impedance is used to identify and classify important characteristics of the transmission system, as explained below. For example, typical characteristic impedances of coaxial cable are 50, 75, and 25 Ω. For two-wire transmission lines, such as TV twin lead, Z_0 is typically 300 Ω.

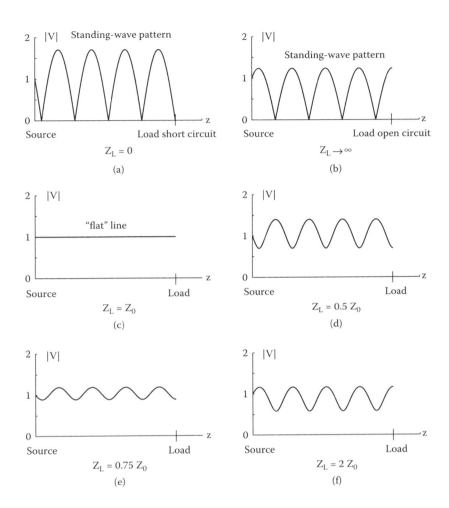

FIGURE 3.30
Envelopes of the voltage as a function of position along the line for various values of load impedance Z_L for the system shown in Figure 3.29. |V| stands for the envelope of V. The voltage source has a magnitude of 1 V. $Z_L = 0$ is equivalent to a short circuit, and $Z_L \to \infty$ is equivalent to an open circuit.

The load impedance has a significant effect on the waves traveling on the transmission system, as illustrated in Figure 3.30, which shows envelopes of the voltage along the transmission system for the configuration of Figure 3.29 for various values of load impedance. A short circuit (perfect conductor connected between the two conductors of the transmission system so $Z_L = 0$, as shown in Figure 3.30(a)) causes a standing wave along the line, exactly like the one in Figure 3.9 for the **E** field of a planewave normally incident on a perfect conductor. An open circuit (zero conductivity between the two conductors so $Z_L \to \infty$, as shown in Figure 3.30(b)) likewise causes a standing wave, but one that is shifted. Other values cause a partial standing wave, but not a pure standing wave, because the minima are not nulls.

An especially interesting situation occurs when $Z_L = Z_0$. For this case, there is no reflected wave, and only a wave traveling to the right exists, as indicated by the envelope in Figure 3.30(c), which is a flat line. When $Z_L = Z_0$, the line is said to be terminated in its characteristic impedance, and the line is also said to be flat (because the envelope is a flat line). Additionally, the line is said to be impedance matched, because the load impedance is equal to the characteristic impedance, or is matched. In cases where a signal or power is to be transmitted to a load, reflections are usually undesirable, and it is usually better to have a matched line.

Two parameters are defined to describe how well a load is matched to a line. One is the magnitude of the reflection coefficient $|\rho|$, which is defined as

$$|\rho| = \frac{|V_{ref}|}{|V_{inc}|} \tag{3.11}$$

where $|V_{ref}|$ is the magnitude of the voltage wave traveling to the left (the reflected wave) and $|V_{inc}|$ is the magnitude of the voltage wave traveling to the right (the incident wave). When $Z_L = Z_0$, the reflected wave is zero, and $|\rho| = 0$. When $Z_L = 0$, $|\rho| = 1$. Also, when Z_L approaches infinity, $|\rho| = 1$.

The second parameter is the voltage standing wave ratio (VSWR), which is defined as

$$VSWR = \frac{|V|_{max}}{|V|_{min}} \tag{3.12}$$

as illustrated in Figure 3.31. In terms of the reflection coefficient, the VWSR is given by

$$VSWR = \frac{1 + |\rho|}{1 - |\rho|} \tag{3.13}$$

Thus, when $|\rho| = 1$, the VSWR approaches infinity, and when $|\rho| = 0$, VSWR = 1. This is consistent with the definition in Equation 3.12, because when $|\rho| = 1$, $|V|_{min} = 0$ (as in Figure 3.30(a)), and when $|\rho| = 0$, $|V|_{min} = |V|_{max}$ (as in Figure 3.30(c)). Thus, in a matched line, VSWR = 1.

Reflections are also produced when transmission lines of different characteristic impedances are connected together. For example, if two striplines having different dielectrics between the two conductors were connected together, the discontinuity in the dielectrics would produce reflections. Figure 3.32 shows a diagram representing the connection of three different transmission systems with different characteristic impedances. The resulting voltage wave patterns for one set of characteristic impedances are shown in Figure 3.33. In general, the voltage partial standing wave patterns are a strong function of the relative characteristic impedances and the lengths of the lines.

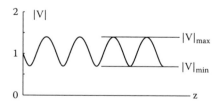

FIGURE 3.31
The maximum and minimum of the voltage envelope used in defining VSWR.

TEM transmission system

FIGURE 3.32
A diagram of three TEM transmission systems, each having a different characteristic impedance and a different length, connected together. A voltage source is at the left end of the combination, and a load impedance is connected to the right end.

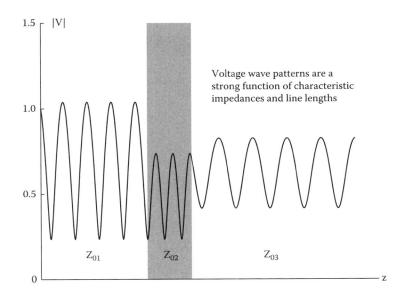

FIGURE 3.33
Voltage wave patterns (partial standing waves) for the configuration of Figure 3.32. The lengths of the lines are 2.2, 1.3, and 2.4 line wavelengths, respectively. $Z_L = 2 Z_{03}$; $Z_{02} = Z_{01}/\sqrt{2}$; $Z_{03} = Z_{01}$.

3.5.2 TEM Systems for Exposing Biological Samples

Some TEM systems, such as the stripline and the parallel-plate waveguide (Figure 3.26), are used for exposing biological samples to **E** and **H** fields. Usually, the exposure system is designed to produce the most uniform **E** fields possible throughout the biological sample. As shown by the **E** field pattern in Figure 3.27 for the stripline, when the dielectric between the conductors is uniform, there is a region in which the **E** field is relatively uniform in a given cross section of the line.

However, if the stripline is suspended in air, and a biological sample is placed between the conductors, the **E** field pattern is modified by the presence of the sample. Figures 3.34 and 3.35 show such patterns for two different instants of time. In Figure 3.34, the **E** fields inside the sample at an instant of time when the source current is zero are negligibly small, and the fields in the surrounding air are relatively very strong. In Figure 3.35, the fields inside and outside the sample are of the same order of magnitude at an instant of time when the source current is maximum, but about four hundred times smaller than

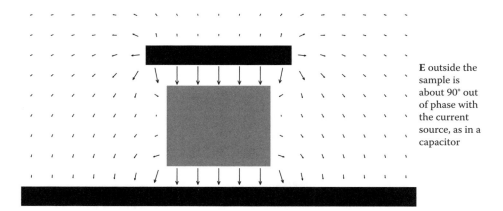

FIGURE 3.34
Two-dimensional simulation of the **E** field pattern when a biological sample is placed between the conductors of the stripline of Figure 3.27, when the conductors are suspended in air. The conductivity and relative permittivity of the biological sample are 0.4 S/m and 100, respectively. The frequency is 10 MHz. The mathematical cells are 1 mm square. The pattern is for an instant of time when the sinusoidal source current is zero.

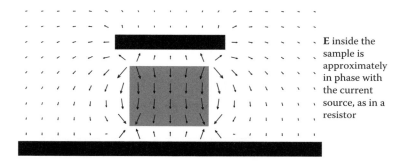

FIGURE 3.35
Same **E** field pattern as in Figure 3.34, except that it is at an instant of time when the sinusoidal source current is maximum. The plotting scale of this figure is about four hundred times greater than in Figure 3.34; that is, the fields in this figure are about four hundred times weaker than those in Figure 3.34.

those in Figure 3.34. These two plots together indicate that the **E** fields outside the sample are almost 90° out of phase with the current source, as in a capacitor, and the **E** fields inside the sample are nearly in phase with the current source, as in a resistor.

In addition to the nonuniformities introduced in the cross-sectional variation of the fields in a lossless biological sample placed in a transmission system, nonuniformities in the direction of propagation also pose a problem when uniform exposure throughout the volume of a biological sample is desired. This problem is illustrated by the examples shown in Figures 3.36 and 3.37. Figure 3.36 shows the voltage wave pattern for the configuration of Figure 3.32, in which the middle characteristic impedance represents the presence of the sample in the transmission system. The variation of |V| with respect to z is directly proportional to the variation of E with respect to z. The length of line 3 and Z_L were chosen so that the |V| in the sample is uniform in the z direction. Theoretically, it is usually possible to do this, but practically, small deviations in the frequency or in the transmission-line dimensions due to temperature changes could upset this balance and cause the |V| in the sample not to be uniform in the z direction.

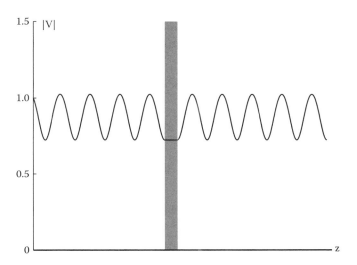

FIGURE 3.36
Voltage wave patterns (partial standing waves) when the middle characteristic impedance of Figure 3.32 represents a lossless biological sample placed in a TEM transmission system.

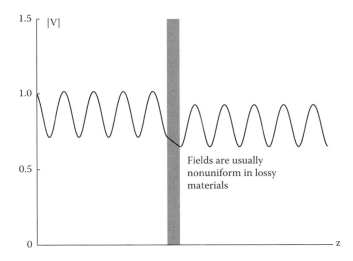

FIGURE 3.37
The same as Figure 3.36 except that the biological sample is lossy.

A more serious limitation is illustrated by the pattern in Figure 3.37, which shows what happens when the biological sample is lossy, as it usually is. The loss in the sample causes |V| to decrease with z. This effect can be minimized to some extent by making the sample as short as possible, but it cannot be entirely eliminated.

Because the **E** fields inside the sample are generally not spatially uniform, and because they are much smaller than those in the surrounding air, careful dosimetry must be carried out to interpret the results of experiments that attempt to relate biological effects to applied **E** fields. Another important consideration is that the simulations described above are based on the existence of the TEM mode alone in a two-dimensional model. At higher frequencies, depending on the size and shape of the sample, other modes can exist

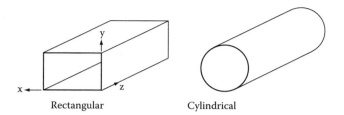

FIGURE 3.38
Two commonly used waveguides, rectangular and cylindrical.

simultaneously. The presence of other modes could cause the **E** fields to be more spatially nonuniform than the examples above indicate.

3.5.3 Waveguides

As stated in the introduction to this section, the three common modes of propagation along guiding structures are TEM, TE, and TM modes. Now that we have discussed TEM mode transmission systems, we are ready to discuss systems in which the modes of propagation are TE and TM modes. These systems are generally called *waveguides*. The most common waveguides are hollow rectangular and cylindrical pipes (Figure 3.38) made of highly conducting material fabricated to close dimensional tolerances.

An electromagnetic source, such as a solid-state microwave oscillator or an electron-beam microwave tube (e.g., a klystron or backward-wave oscillator), is often connected to one end of the waveguide, which produces EM waves that propagate down the waveguide. The other end of the waveguide might be connected to a radar antenna, such as a dish, to transmit signals out into space. Or a microwave receiving antenna might be connected to one end of the waveguide and the waveguide used to transmit the received signal to a microwave receiver connected to its other end.

Waveguides are also used to expose biological samples to EM fields. When biological samples are inserted into waveguides, the same considerations with respect to uniformity of the **E** and **H** fields inside the sample occur as with the TEM systems, as described in Section 3.5.2.

To illustrate the concepts of TE and TM modes and their propagation in waveguides, we shall discuss in detail these modes in rectangular waveguides. Similar properties and characteristics apply to other kinds of waveguides, such as cylindrical waveguides.

3.5.3.1 TE and TM Mode Patterns in Rectangular Waveguides

An infinite number of TE and TM modes can exist in a rectangular waveguide. Each one consists of a characteristic combination of **E** and **H** field distributions. These modes are designated as TE_{mn} and TM_{mn} modes, where m and n are digits that identify each of the modes. In general, an infinite number of these TE and TM modes can be present simultaneously in a waveguide, depending on the method of excitation, the size and shape of the waveguide, and the frequency of the waves, as will be explained later. The variable m specifies how many peaks occur in the mode pattern across the width of the waveguide (the x direction in Figure 3.38), and the n specifies how many peaks occur in the mode pattern across the height of the waveguide (the y direction in Figure 3.38).

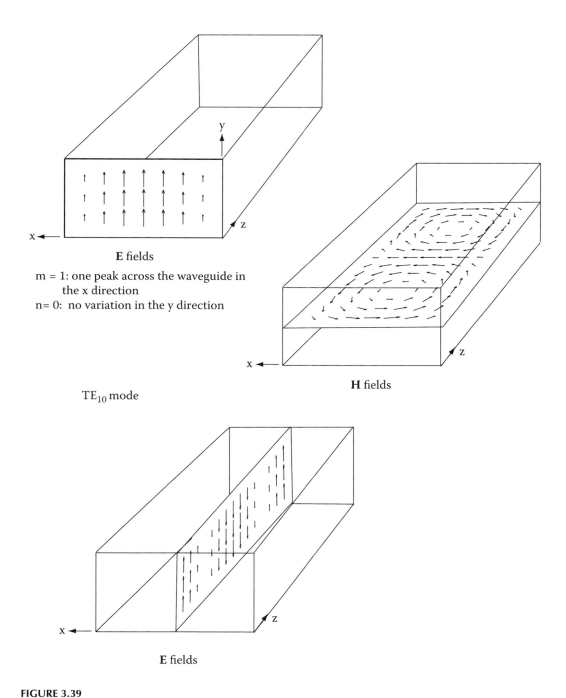

E fields

m = 1: one peak across the waveguide in the x direction

n= 0: no variation in the y direction

TE_{10} mode

H fields

E fields

FIGURE 3.39
The TE_{10} mode patterns in three different planes in a rectangular waveguide at one instant of time. As time progresses, the patterns move in the +z direction.

Figure 3.39 shows the **E** and **H** vector field patterns at one instant of time for the TE_{10} mode, in which m = 1 and n = 0. The pattern propagates down the waveguide as time goes on. m = 1 indicates that there is one peak across the waveguide. This corresponds to the E_y

(the component of **E** in the y direction) having a maximum in the center and being zero on each side wall. Note that in all cases, the boundary conditions (see Section 1.12) require that the tangential component of the **E** field be zero at the metallic walls (assuming that they are perfectly conducting) of the waveguide. Or, in other words, the **E** field must be normal to the perfectly conducting walls. Because E_y is tangential to the side walls, it must be zero there, but because it is normal to the top and bottom walls, it need not be zero there.

n = 0 indicates that there is no variation in the y direction; that is, at a given value of x, all three vectors shown in the diagram have the same length in y. As another example, patterns for the TE_{20} mode are shown in Figure 3.40. Again, there is no variation in the y direction, but m = 2 indicates that two peaks occur across the waveguide. In this case, E_y is zero in the center as well as at the two side walls.

The peaks in the mode patterns are more clearly displayed in terms of the envelopes (see discussion in connection with Figure 3.9 in Section 3.3.1) of **E**, as shown in Figure 3.41 for three modes, TE_{10}, TE_{20}, and TE_{11}. The TE_{10} mode has one peak across the waveguide, and the TE_{20} mode has two peaks across the waveguide. The TE_{11} mode pattern is more complicated than the other two in Figure 3.41 because in it there is one peak across the guide in both the x direction and the y direction, as shown in Figure 3.41(c). Again, the **E** fields are normal to the metallic walls at the walls.

These examples illustrate the nature of mode patterns. Other patterns in rectangular waveguide are similar in nature and behavior. Mode patterns in other kinds of waveguide,

RESONANCE EFFECTS IN THE HUMAN BODY

The resonance phenomenon also describes waves that partially reflect back and forth in some sections of the body, "reverberating" until they are absorbed rather than simply passing through the body. This causes much more power to be absorbed at specific frequencies than would be predicted simply using an analysis of attenuation. Attenuation effects alone usually absorb more power at high frequencies than low frequencies, but resonance causes the body to absorb higher power within a small band of frequencies, with ranges on either side that have less absorption. Prediction of the whole-body resonance can be approximated reasonably well by treating the body as a half-wave dipole (where the body height is one-half of a wavelength) if the feet are ungrounded, or as a monopole (where the height is one-quarter of a wavelength, due to the presence of a ground plane) if the feet are grounded. Thus, a 6-foot-tall human should resonate around 75 MHz if ungrounded and 38 MHz if grounded. This effect is indeed seen.

This does not tell the whole story, though. Individual parts of the body, such as the torso, can resonate at higher frequencies than the whole body. Also, animals of different sizes and shapes will have different resonant frequencies than a human, for instance. This can cause some difficulties when doing biological tests of electromagnetic fields. If a rat is being treated at its resonant frequency, the dose of deposited electromagnetic power at that frequency can be far higher than it would be in a human (who is not resonant at the same frequency). It is nontrivial but important to assess these effects.

Thus, electromagnetic absorption in humans is strongly dependent on frequency and is controlled not only by the properties of the tissue, but also by the shape and such as the cylindrical waveguide, are similar in characteristic behavior, but differ in detail because of the round shape of the waveguide. Now that we have discussed what modes are, we next discuss how they are excited and how they exist in combinations in waveguides.

3.5.3.2 Mode Excitation and Cutoff Frequencies

Waveguides are typically excited by a microwave generator, such as an electron-beam tube like a magnetron, or a solid-state device like a transistor oscillator, that is connected to the waveguide by a coaxial cable. The coaxial cable is typically connected to the waveguide by extending the center conductor of the cable through the top waveguide wall to form a probe in the waveguide, and connecting the outer conductor of the cable to the waveguide wall, as shown diagrammatically in Figure 3.42. The center wire of the coaxial cable can also be formed into a loop and connected back to the waveguide wall to produce loop excitation. To excite the TE_{10} mode, the coaxial cable is usually put in the side wall of the waveguide so that the plane of the loop is perpendicular to the **H** fields. The probe, or whatever other method of introducing the microwave fields into the waveguide is used, excites many TE and TM modes that exist simultaneously in the waveguide. Some of these modes propagate, however, and some die away very rapidly from the probe. The modes that die away are called *evanescent* modes.

---

size of the body. Natural body resonances, where the body or parts of the body absorb significantly higher amounts of energy than would be predicted simply from physical cross sections, must be taken into account when setting safety guidelines.

The resonant frequencies in the human body and head are shown in Table 3.1 for ungrounded planewave exposure conditions. The absorption cross section is nearly three times larger than the physical cross section for the head, indicating that at this resonant frequency nearly three times as much power is absorbed in the head as would be predicted from the physical cross section. Resonance typically scales with the size of the object relative to wavelength. So smaller objects (such as a child's head compared to an adult's head) typically have resonances at higher frequencies. Also, whole-body resonance is seen at a much lower frequency than part-body resonances, such as in the head.

TABLE 3.1

Whole-Body and Head Resonance Effects for Ungrounded Humans

	Head Resonant Frequency (MHz)	Head Absorption Cross Section/Physical Cross Section	Whole-Body Resonant Frequency (MHz)—Ungrounded	Whole-Body Absorption Cross Section/Physical Cross Section
Adult	205	2.78	75	3.92
10-year-old	270	2.84	100	3.88
5-year-old	330	2.92	130	3.47

Source: Furse, C., et al., Conditions for resonant absorption in the human head for planewave exposure, paper presented at 2nd World Congress for Electricity and Magnetism in Biology and Medicine, Bologna, Italy, June 8–13, 1997. With permission.

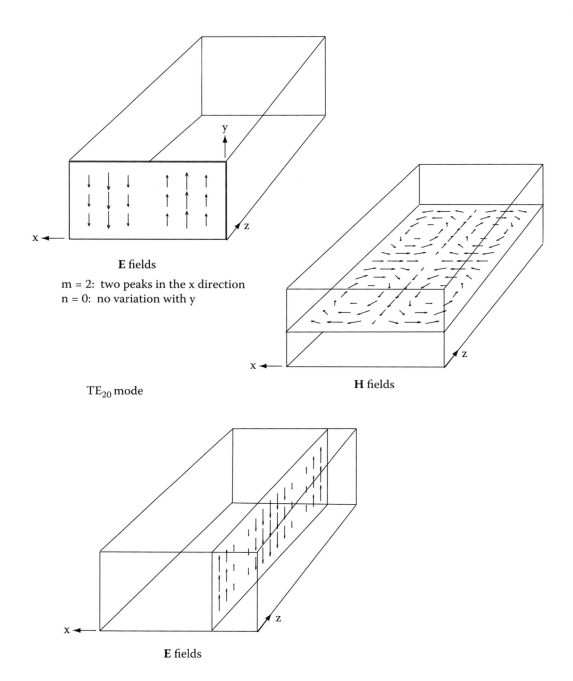

FIGURE 3.40

The TE_{20} mode patterns in three different planes in a rectangular waveguide at one instant of time. As time progresses, the patterns move in the +z direction.

Whether a mode propagates or is evanescent is determined by the frequency of the exciting fields and the size and shape of the waveguide. For each mode, there is a frequency, called the *cutoff frequency* (usually designated by f_{co}) below which the mode will be evanescent and above which the mode will be propagating. The cutoff frequency for each mode

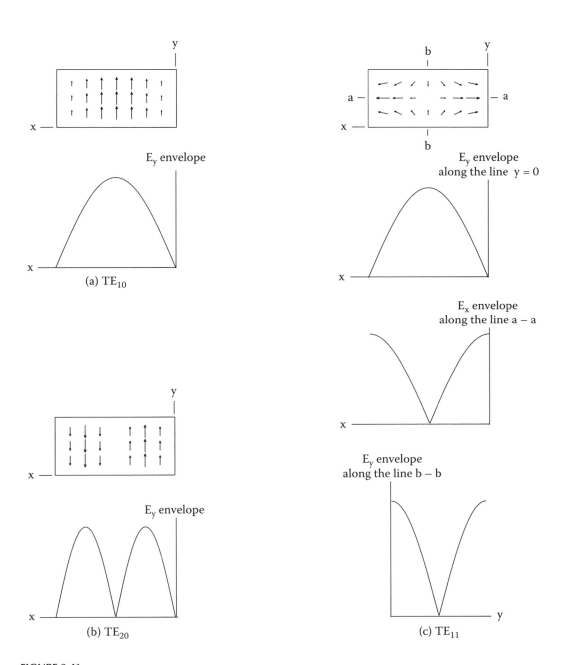

FIGURE 3.41
The mode patterns in one cross section of rectangular waveguide at one instant of time, and corresponding envelopes of **E**, for the TE_{10}, TE_{20}, and TE_{11} modes. The envelopes in (a) and (b) are the same for any horizontal line across the waveguide because in the TE_{10} and TE_{20} modes there is no variation with y. In (c), the envelopes are along the lines indicated.

depends on the size and shape of the waveguide. For rectangular waveguide, the TE_{10} mode has the lowest cutoff frequency. Figure 3.43 shows a diagram of how the cutoff frequencies of the various modes are related when b = a/2, where b is the height and a is the

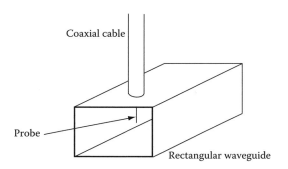

FIGURE 3.42
Diagram of a method of coupling from coaxial cable into rectangular waveguide. The probe excites **E** fields in the waveguide.

FIGURE 3.43
Cutoff frequencies for various modes normalized to the cutoff frequency of the TE_{10} mode in a rectangular waveguide in which the height is one-half the width.

width of the waveguide. $b = a/2$ is the condition for which the greatest separation between the TE_{10} and the next modes occurs.

Usually conditions are chosen so that only one mode will propagate in the waveguide, with all the rest of the modes being evanescent. The frequency is therefore adjusted so that it is above the cutoff frequency for the mode with the lowest cutoff frequency (TE_{10} in Figure 3.43), but below the cutoff frequency of the next modes (TE_{01} and TE_{20} in Figure 3.43). Then many evanescent modes may exist immediately around and near the exciting probe and near any discontinuities in the waveguide, but in a smooth and regular waveguide, only the propagating mode will exist for any appreciable length. This situation is usually desirable because a combination of propagating modes with different amplitudes and different velocities of propagation produce complicated field patterns that are difficult to implement and control.

The cutoff frequency for the TE_{10} mode occurs when $a = \lambda_{co}/2$, where λ_{co} is the free-space wavelength at the cutoff frequency. Because wavelength is inversely proportional to frequency (see Equation 1.15), a smaller waveguide is used at higher frequencies. For example, if $f = 10$ GHz and it is desired that this frequency be halfway between f_{co} for the TE_{10} mode and the next higher mode (see Figure 3.43), then f_{co} for the TE_{10} mode should be 6.67 GHz, for which λ_{co} is given by $\lambda_{co} = 3 \times 10^8/6.67 \times 10^9 = 0.045$ m. Thus, the waveguide should be 0.0225 m wide and 0.01125 m high (remember, $b = a/2$ in Figure 3.43). For an operating frequency of 20 GHz, the waveguide would be half as large.

3.5.3.3 Waveguide Systems for Exposing Biological Samples

Samples are sometimes placed in various kinds of nonmetallic containers in waveguides to expose them to microwave fields. Considerations of such systems are similar to those discussed in Section 3.5.2 for biological samples placed in TEM systems. The mode patterns in the cross section of a waveguide, though, can be considerably different from those in the cross section of TEM systems.

The lossiness of the sample, however, causes the same kind of effects in waveguide exposure systems as those illustrated in Figures 3.36 and 3.37 in connection with TEM exposure systems. In those figures, the magnitude of the voltage is shown as a function of distance in the direction of propagation. In a waveguide exposure system, similar effects apply to the E field. It is often difficult to ensure that a biological sample in a waveguide is exposed to uniform fields throughout the sample.

3.6 Resonant Systems

Resonance is an effect that is important in the frequency range treated in this chapter. The basic phenomenon of resonance is illustrated by the excitation of the two-dimensional model of a cavity shown in Figure 3.44. A cavity is a hollow enclosure in which EM fields can be excited. Commonly used cavities consist of a section of waveguide, either rectangular or cylindrical, with conducting walls added at each end. EM fields are excited inside the cavity by a probe or loop or some other connection to a source. In the two-dimensional model of Figure 3.44, a current source is connected across the hole in the left wall of the cavity.

The strength of the E and H fields excited in the cavity is a strong function of the frequency of the source and the size and shape of the cavity. Figure 3.44 shows the E fields at one instant of time at the lowest frequency for which the E fields are strongly excited (keeping the current source magnitude constant). Figure 3.45 shows the response of the cavity as a function of the frequency of the current source when the inside dimensions of the cavity are 23 × 23 cm. For the purposes of this illustration, we define the response of the cavity as the sum of the squares of all the E fields inside the cavity, which is proportional to the energy stored in the E fields inside the cavity for a fixed source strength. A strong resonance occurs at a frequency of 670.2484 MHz. That is, the response is much stronger at that frequency than at other adjacent frequencies.

Figure 3.46 shows the effects of adding some slightly lossy material to the cavity. The loss in the cavity makes the response curve wider and lower. The relative width of the response curve is called the *bandwidth*. The lossy curve has a wider bandwidth than the lossless curve. When the bandwidth is narrow, the Q of the cavity (Q stands for quality factor) is higher than when the bandwidth is wider. Thus, the lossy cavity has a lower Q and a wider bandwidth than the lossless cavity.

Resonance occurs near frequencies for which multiples of half wavelengths fit across the cavity from left to right or from top to bottom. The lowest resonant frequency occurs near the frequency at which one half wavelength occurs from left to right and zero half wavelengths from top to bottom (i.e., the field does not change from top to bottom). The frequency for which one half wavelength fits exactly across the cavity is found from Equation 1.15. For $\lambda/2 = 23$ cm, $\lambda = 46$ cm, and $f = c/\lambda = 3 \times 10^8/46 \times 10^{-2} = 652.1739$ MHz. When one half wavelength fits exactly across the cavity, a null occurs at the left wall and a null at the right wall, with an envelope like that shown in Figure 3.41(a). However, according to the finite-difference frequency-domain (FDTD) numerical calculations for the two-dimensional model shown in Figure 3.44, the resonant frequency is 670.2484 MHz, which is about 3% higher than 652.1739 MHz. The reason the resonant frequency is slightly higher than 652.1739 MHz is that the hole in the wall perturbs the pattern slightly, causing a null just to the right of the hole, as shown in Figure 3.44.

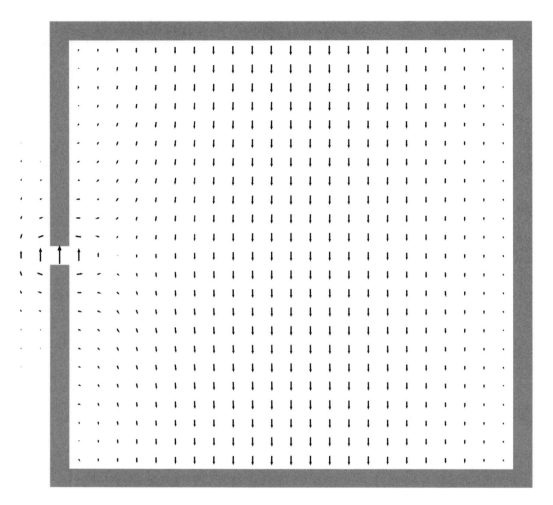

FIGURE 3.44
E fields in a two-dimensional model of an air-filled cavity excited by a current source (not shown) connected across the gap in the left cavity wall with a frequency of 670.2484 MHz. The inside dimensions of the cavity walls are 23 × 23 cm.

Figure 3.47 shows the **E** field pattern for a higher-order mode at a resonant frequency of 1,315.9489 MHz. This pattern shows variation in the fields both from left to right and from top to bottom. Many other resonant frequencies exist for this two-dimensional model with multiple variations in both directions, across and up and down. In actual three-dimensional cavities, multiple resonant frequencies occur with multiple variations in all three directions across the cavity. Similar effects occur in cylindrical cavities and cavities of other shapes.

The simple two-dimensional model described above illustrates the general characteristics of resonance. EM fields are strongly excited in cavities at very specific frequencies; at other frequencies, the EM fields inside the cavity are only very weakly excited. Cavities are used in many applications where frequency discrimination is required, such as in tuners

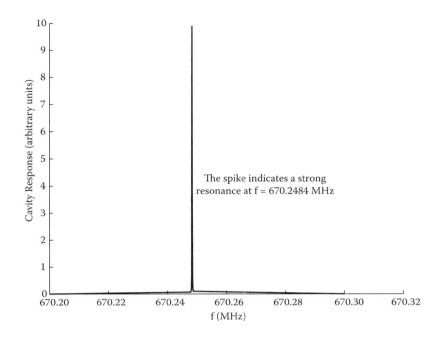

FIGURE 3.45
Cavity response as a function of excitation frequency for the cavity of Figure 3.44. Cavity response is the sum of the squares of all the **E** fields inside the cavity, which is proportional to the energy stored in the **E** fields. A strong resonance occurs at 670.2484 MHz.

for selecting a signal of a specific frequency out of multiple signals of various frequencies, and in filters that reject signals as a function of frequency.

Biological samples are also sometimes placed in cavities to expose them to EM fields. As illustrated by the example above, however, the loss in the samples will considerably lower the Q of the cavity, lower the fields inside the cavity, and increase the bandwidth. Furthermore, because the **E** and **H** fields in the cavity are generally not very uniform, care must be taken in determining the dosimetry of the EM fields in the sample.

Figure 3.48 shows the cavity response when a biological sample having a conductivity of 0.6 S/m and a relative permittivity of 100 is placed in the center of the cavity. Note from the figure caption that the cavity response in Figure 3.48 is one million times weaker than that shown in Figure 3.45 for the empty cavity. The presence of the sample has changed the resonant frequency markedly, as well as lowering the Q and increasing the bandwidth. Figure 3.49 shows the **E** fields inside the cavity with a biological sample placed at its center at the resonant frequency of 644.937 MHz. The pattern shows a perturbation of the pattern of the empty cavity, with the **E** fields inside the sample much weaker than those outside the sample, which we would expect because the conductivity and relative permittivity of the sample are both relatively high. Also, the phase of the **E** fields in Figure 3.49 relative to the current source is about 90° offset from the phase of the **E** fields in the empty cavity relative to the current source. This phase difference is caused by the high relative permittivity of the sample in the cavity. An understanding of resonance effects is obviously important in designing experiments and interpreting results.

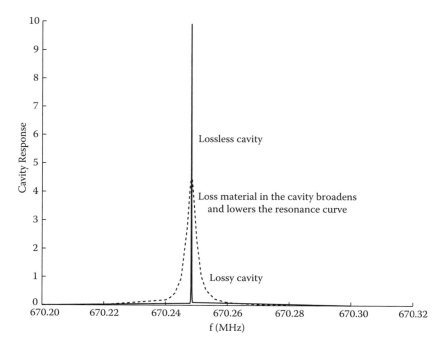

FIGURE 3.46
Cavity responses for the cavity of Figure 3.44 for an air-filled cavity (lossless) and for a cavity filled with a material having a conductivity of 2×10^{-7} S/m (lossy). The response for the lossy cavity has been multiplied by 100 to make the graph visible on the same set of axes as that of the lossless cavity.

3.7 Antennas

As explained in Section 1.17, when the wavelength is on the order of the size of the system, energy can be efficiently beamed through the air, as well as being transmitted through coaxial cables and waveguides. A typical system for transmitting EM signals through the air consists of a source, such as a radio transmitter or microwave generator, a transmission line or waveguide, and an antenna. The source produces EM fields that propagate along the transmission line or waveguide to the antenna, which launches the propagating wave into space, where it propagates similarly to how the waves described in the first part of this chapter propagate. Such antennas are called *transmitting antennas*. Antennas are also used to receive EM radiation,* which is then propagated along a transmission line to a receiver. These antennas are called *receiving antennas*. An antenna can generally be used as either a transmitting antenna or a receiving antenna. In this section, we describe some of the general properties and characteristics of antennas.

Antennas are classified into several groups: wire antennas, aperture antennas, array antennas, reflector antennas, and lens antennas. Wire antennas are various combinations of wires or rods. Some commonly used ones are shown in the upper part of Figure 3.50. A dipole antenna consists of two segments of rod or wire, with a transmission line connected

---

* *Radiation* is a term often used to describe an EM wave propagating through space, or "radiating." EM waves are very different from x-ray radiation, however, because they do not ionize biological materials.

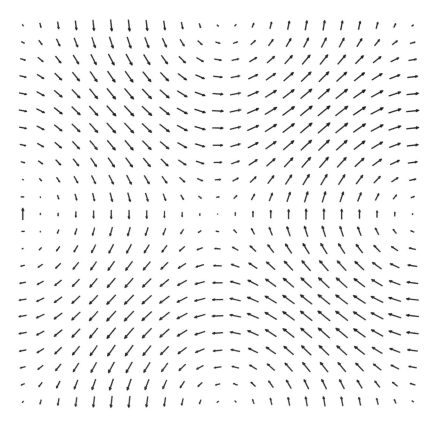

FIGURE 3.47
E fields inside the air-filled cavity of Figure 3.44 when the excitation is 1315.9489 MHz.

between them. The length of a dipole antenna is typically one-half of a wavelength. A folded dipole, as the name indicates, is a dipole with an additional connection between the ends. Loop antennas may be circular, square, or other shapes.

Aperture antennas are openings through which EM waves are launched into space, such as horns—like the pyramidal horn shown in the lower part of Figure 3.50—open-ended waveguides, slots in waveguides, or other kinds of apertures. Array antennas are assemblies of various kinds of antennas, such as a series of dipole antennas. TV receiving antennas are usually array antennas, with one of the rods being the active, or driven, element, and the other rods serving as directors or reflectors of EM radiation. Open-ended waveguides are sometimes used to deliver EM power in a localized way for hyperthermia for cancer therapy. Once the wave has left the waveguide, however, it will spread out (diffract) and attenuate as it passes into the body. For this reason, waveguides are generally only used to treat superficial tumors.

A typical reflector antenna is the parabolic microwave dish commonly used to receive TV signals from satellites. The parabolic reflector concentrates the microwave energy at its focal point. A visual example of how parabolic reflectors work can be seen in the large parabolic mirrors in operating room lights or the light a dentist typically uses. Lens antennas are used to form EM radiation into beams for transmission into space, or to receive and concentrate EM radiation. Because lens antennas must be large compared to a wavelength

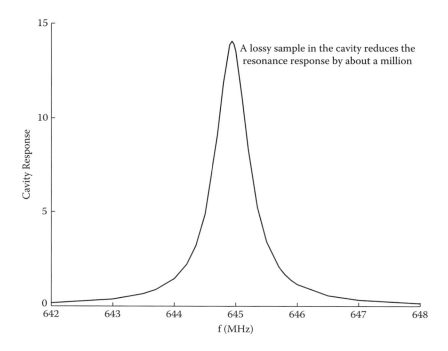

FIGURE 3.48
Cavity response for the cavity of Figure 3.44 in which a biological sample with a conductivity of 0.6 S/m and a relative permittivity of 100 has been placed. This response is about a million times weaker than that of Figure 3.45 (the relative scale here differs from the one in Figure 3.45 by a factor of a million).

if they are to be effective, they are used primarily at higher frequencies where wavelengths are smaller.

Directional antennas transmit or receive radiation more effectively in some directions than others. A laser pointer is an example of a directional source. *Isotropic antennas* transmit or receive radiation equally in all directions. A lightbulb is close to an isotropic source. Only ideal antennas (e.g., a point source), not physically realizable antennas, are truly isotropic. *Radiation patterns* are used to describe the characteristics of antenna radiation or reception. Radiation patterns can be either *field patterns* or *power patterns*. Field patterns show either the **E** field or the **H** field as a function of position, and power patterns show the power as a function of position. Fields near the antenna are called *near fields*, and fields farther away from the antenna are called *far fields*. The distance from the center of the antenna to where the far fields begin is given by

$$R = 2D^2/\lambda \qquad (3.14)$$

where D is the largest dimension of the antenna and λ is the wavelength. The near fields vary more rapidly with space than the far fields. **E** and **H** are not necessarily perpendicular in the near fields, and the near fields are not so much like propagating waves. In the far fields, **E** and **H** are perpendicular, and the fields have the characteristics of propagating waves.

Figure 3.51 shows the far-field **E** field pattern for a thin half-wavelength dipole. The plot is called a polar plot. In a polar plot, each point in the plane is located by two coordinates,

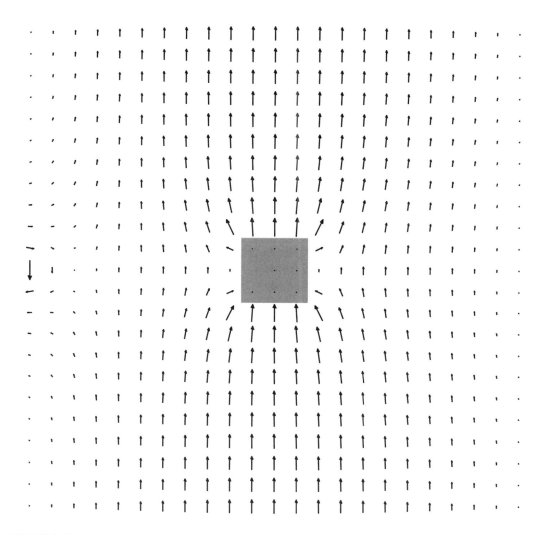

FIGURE 3.49
E fields in the cavity of Figure 3.48 at the resonant frequency of 644.937 MHz.

the distance out from the origin (center), and the angle from a vertical line. In the radiation pattern, the angle of each point on the curve is the angle at which the magnitude of the E field is calculated, and the distance out from the origin represents the relative magnitude of the E field at that given angle. The magnitudes of the E field are all calculated at a given far distance from the antenna, but that distance does not show up on the plot since in the far fields, the angular dependence of E does not depend on the distance away from the antenna.

In any plane containing the antenna of Figure 3.51, the radiation pattern is the same. Thus, looking toward the end of the antenna, the radiation pattern would be a circle centered about the antenna. In a three-dimensional representation, the radiation pattern would appear as a donut around the antenna. No power comes out the end of a wire antenna. As shown by Figure 3.51, the maximum radiation occurs broadside to a dipole antenna, that is, at 90° to the antenna. Furthermore, the antenna does not radiate off its

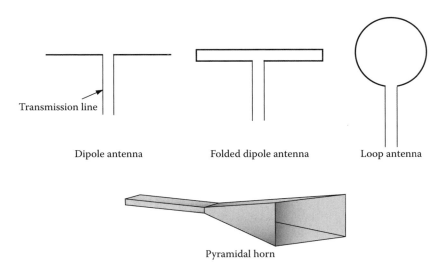

Transmission line

Dipole antenna   Folded dipole antenna   Loop antenna

Pyramidal horn

FIGURE 3.50
Examples of antennas. The top three are called wire antennas, and the pyramidal horn is an example of an aperture antenna.

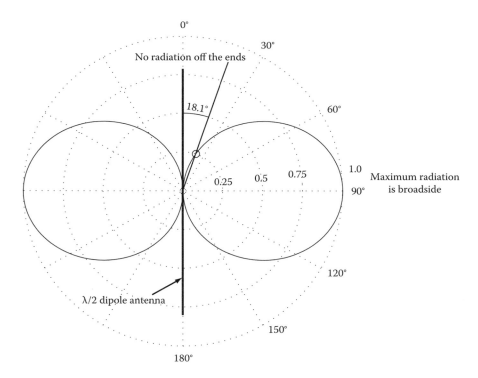

FIGURE 3.51
E field radiation pattern of a half-wavelength dipole antenna.

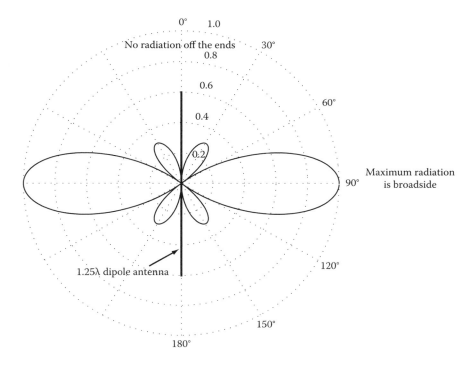

FIGURE 3.52
E field radiation pattern of a 1.25-wavelength dipole antenna.

ends, as indicated by the nulls in the pattern at 0° and 180°. This is a general characteristic of wire antennas; they do not radiate off their ends.

As another example of this characteristic, Figure 3.52 shows the radiation pattern for a dipole antenna that is 1.25 wavelengths long. Again, the pattern shows that no radiation occurs off the end. The pattern also shows an additional effect not seen in Figure 3.51, the presence of *major lobes* and *minor lobes*, or *side lobes*. In some applications, radiation in one direction only is desirable. For example, in transmitting from a fixed radio transmitter to a fixed radio receiver, radiation in other directions is essentially wasted. In these cases, the antenna should be very directional, having one large, narrow lobe and no side lobes. This ideal situation is not usually attainable, but some antennas come close to it.

Dipole antennas are seen only rarely in medical applications, because they require two arms of equal length. More commonly, one arm of a linear antenna (a wire) will be placed above a conducting ground plane, creating a monopole antenna. The ground plane creates a virtual reflection below the plane, making the monopole appear electrically like a dipole antenna, as shown in Figure 3.53. The shape of the top half of the radiation pattern for a monopole is identical to a dipole, but since a dipole sends the power into both the top and bottom halfspaces, while the monopole sends it only into the top halfspace, if both antennas were fed with equal power, the dipole antenna would radiate half as much in any given direction. Also, the feedpoint electrical impedance is half as much for a monopole as a dipole antenna. Monopole antennas are used to deliver localized power, and therefore to heat very focused regions of the body for applications such as cardiac ablation and hyperthermia for cancer therapy. They are also used for measuring the electrical proper-

FIGURE 3.53
Monopole antenna on a cellular telephone. The antenna is fed at the base as shown by the circle. The part of the antenna above the base is the actual radiating arm of the monopole. The telephone case is metalized on the inside and acts like a ground plane (albeit a very imperfect one). A virtual image of the antenna is created as shown. The effect is that the monopole antenna above the ground plane acts very much like a dipole antenna.

ties of materials. As shown in the patterns of Figures 3.51 and 3.52, dipole and monopole antennas are considered to be between directional and isotropic.

The microstrip patch antennas shown in Figure 3.54 are close to being isotropic antennas in the region above the patch. These antennas are made from two plates of metal (the top one of various shapes) with a low-loss dielectric material sandwiched between them. A coaxial cable is extended through a hole in the bottom metal plate (the ground plane). The center conductor of the coax is attached to the top patch (typically near one edge), and the ground shield of the coax is connected to the bottom plate. Care is taken to prevent touching the inner conductor of the coax to the bottom plate. Alternatively, the coax feeds a microstripline as shown in Figure 3.54. The location of this feed system is very critical to the impedance matching of the antenna to the source. Regardless of how odd the shape of the patch antenna, the radiation pattern stays very similar, although the input impedance may change significantly from one design to another. Patch and other plate

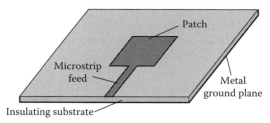

FIGURE 3.54
Microstrip patch antenna. An insulating substrate (typically a few millimeters thick) has a metal ground plane below and a printed (metal) patch above. The feed is a microstripline, typically centered on one side of the patch. Many different shapes of patch are available today, as well as many different methods of feeding these patch antennas.

antennas radiate above the patch but do not radiate below the patch where the ground plane is located, but two patch antennas back-to-back can provide near-isotropic coverage. Applications are emerging in biomedical telemetry, which entails communication between an implantable or body-worn medical device and an external receiving antenna.

Directional antennas are used to focus energy for hyperthermia for cancer therapy and for MRI scans of localized regions of the body. In general, an antenna must be large in order to focus energy, or several antennas must be used in an array. For an array, several antennas are fed at once, and the waves are slightly out of phase due to the difference in their locations and (sometimes) the phases with which they are fed. The wave from each antenna in the array adds (in a phasor fashion) to all of the others. The constructive and destructive interference of these combined waves produces peaks (focused locations) and nulls (zero locations) of the waves, just as it does for the reflected waves in front of a metal halfspace. Directional arrays are often used for whole-body hyperthermia heating and for medical imaging applications.

In bioelectromagnetic dosimetry research, both directional and isotropic antennas are used. In some, a directional antenna is used to expose one animal or one sample. In other applications, antennas with less directionality are used to expose many animals. A basic understanding of the characteristics of antennas is important in designing and interpreting experiments in which antennas are used to expose biological specimens, especially in ensuring satisfactory dosimetry.

MICROWAVE TELEMETRY

Today's standard for communicating with implantable medical devices (implantable cardiac pacemakers and defibrillators, deep brain stimulators for treatment of Parkinson's disease, and implantable hormone therapy pumps, for example) utilizes inductive telemetry, as described in Section 1.5. This method requires that the external receiving coil be within a few inches of the body, and only a limited data rate can be transferred. Typical frequencies of operation for inductive telemetry are under 10 MHz.

Recently, a Medical Implant Communication Services (MICS) band was created from 402 to 405 MHz for communication with long-term biocompatible devices. Limits on its bandwidth and power and the fact that MICS shares the frequency allocation with the Meteorological Aids Service (METAIDS), which is used primarily by weather balloons, limit this band to indoor use. This band was chosen because it is low enough in frequency that a reasonable amount of power can be transferred into the body (see Figure 1.40), yet high enough that the wavelength is sufficiently small that efficient implantable antennas can be built for this application. For instance, the wavelength in tissue with properties two-thirds that of muscle (a common approximation for the average body) at 433 MHz is 10.1 cm, and 16.8% of the field will remain after the wave has traveled 10 cm through the body. (See the example at the end of Section 1.14.) Communication systems for this frequency range are still in the early stages of development; however, sufficient success has been achieved that it appears likely these systems will become the industry standard for communication with implantable devices in the future, thus "untethering" the patient from the receiver and opening up many new applications in telemedicine and patient monitoring. Power delivery (for recharging batteries) will still require inductive links. *(continued on next page)*

Designing an implantable antenna for use in the body has some unique challenges. First, the antenna must be small enough to fit in the body. With a wavelength of 10.1 cm, a typical dipole antenna would have to be half of this wavelength, or 5.05 cm. This is usually too long to place in the body, although some groups have experimented with using the various metallic leads already incorporated in the medical device (particularly for cardiac pacemakers) to act as the antenna. Another alternative is to use microstrip antennas such as those discussed in Section 3.7. The microstrip antenna uses a trick to reduce its required dimensions. The dielectric material of the insulating substrate will reduce the apparent wavelength on the antenna, and thus reduce the size of antenna required. The approximate size reduction is proportional to $\sqrt{\varepsilon_r}$ of the substrate material. Typical values of $\varepsilon_r = 4$–10 are available, although materials up to 100 are available for other (nonbiocompatible) applications. This reduction in size brings the microstrip into the range of acceptable antenna designs (but not without a cost: these antennas are relatively inefficient).

Spiral, serpentine, and waffle-like microstrip antennas have been developed for this application. The spiral and serpentine designs (Figure 3.55 and Figure 3.56) act like a cross between monopole antennas (note the ground pin that creates the virtual image plane) and inductive coils. The waffle antenna (Figure 3.57) is entirely different. This antenna is designed with a genetic algorithm that takes a series of random designs and combines them in ways resembling natural selection to arrive at a best possible design.

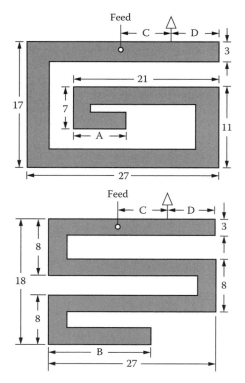

FIGURE 3.55
The (top) spiral and (bottom) sepentine microstrip antennas (units in mm). (From Soontornpipit, P., et al., *IEEE Trans. MTT*, 52, 1944–51, 2004. © 2004 IEEE. With permission.)

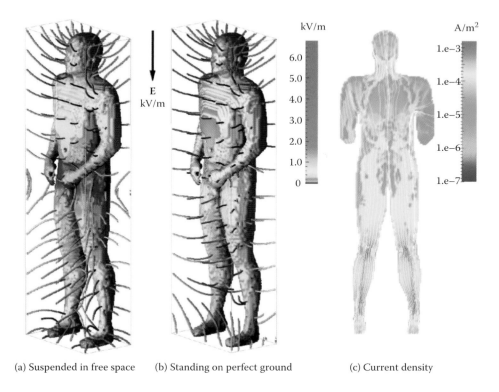

(a) Suspended in free space   (b) Standing on perfect ground   (c) Current density

COLOR FIGURE 2.39
The electric field vectors intersecting a human body standing in a uniform 1 kV/m vertical electric field at 60 Hz for the case of (a) insulated feet and (b) grounded feet. The current density in a vertical cross section of the body is shown in (c). (From Stuchly, M., and Dawson, T., *Proc. IEEE*, 88, 643–64, 2000. © 2000 IEEE. With permission.)

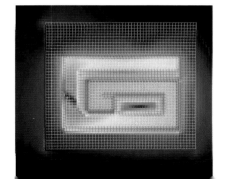
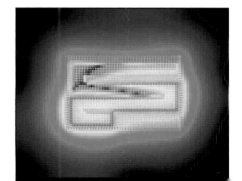

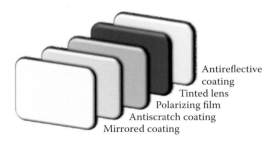

COLOR FIGURE 4.10
Layers commonly found on sunglass lenses.

Antireflective
coating
Tinted lens
Polarizing film
Antiscratch coating
Mirrored coating

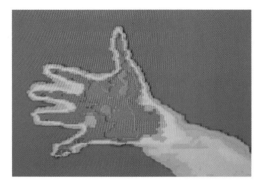

COLOR FIGURE 4.26
Infrared image of a hand taken with a thermographic camera. (From M. Iskander. With permission.)

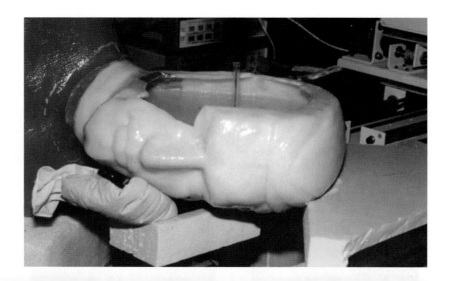

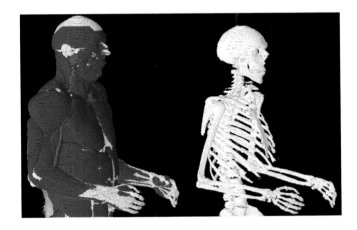

COLOR FIGURE 5.4
The visible man from the National Institutes of Health was originally scanned with the arms at his sides. Here they have been computationally repositioned, with care being taken to ensure that the bones and other structures remain intact. (From Remcom, Inc. Reprinted with permission.)

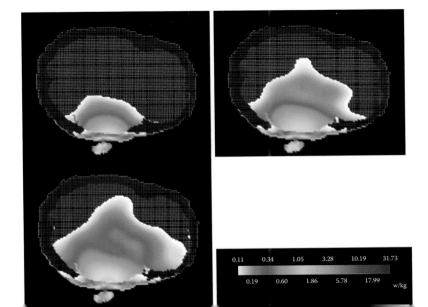

| 0.11 | 0.34 | 1.05 | 3.28 | 10.19 | 31.73 |
| 0.19 | 0.60 | 1.86 | 5.78 | 17.99 | w/kg |

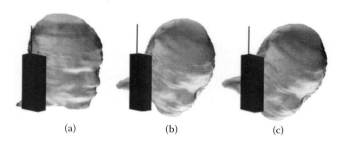

(a)   (b)   (c)

COLOR FIGURE 5.11
Orientation of an 835 MHz cellular telephone near the head.

COLOR FIGURE 6.11
A millimeter-resolution NMR coil wrapped around a needle. The needle and coil are then encapsulated in plastic or rubber, and the needle is removed. This creates a 100% fill-factor coil. (From J. Stephenson.)

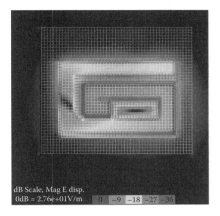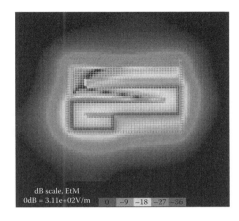

FIGURE 3.56
Please see color insert following page 146. Distributions of the magnitude of **E** on spiral and serpentine antennas designed for communication with cardiac pacemakers. (From Soontornpipit, P., et al., *IEEE Trans. Antennas Propagation*, 53, 1939–45, 2005. © 2005 IEEE. With permission.)

Ground Feed

FIGURE 3.57
Waffle antenna designed with a genetic algorithm. (From Soontornpipit, P., et al., *IEEE Trans. Antennas Propagation*, 53, 1939–45, 2005. © 2005 IEEE. With permission.)

MICROWAVE IMAGING

Originally it was assumed that microwave imaging would be highly problematic and would never achieve the results seen in x-ray or MRI imaging, due to the relatively long wavelengths at frequencies that can penetrate through the body. This is certainly true from the point of view of trying to focus the fields using hardware alone; however (like MRI), the use of advanced image processing methods has significantly enhanced microwave imaging to the point that it is receiving significant attention as a promising imaging method.

At microwave frequencies, the wavelengths are shorter than for low-frequency imaging such as impedance imaging (Section 2.5.1.3). Thus, the waves can be more directed through the body, avoiding the problems of the chest wall shunting currents away from the interior that occurs in impedance imaging. Microwave images have been obtained by either rotating the object or rotating the *(continued on next page)* microwave applicator and receiver, or by using an array of microwave applicators and receivers where the field is electrically scanned rather than rotated.

Although microwave imaging has some advantages over impedance imaging, other difficulties are encountered in trying to make microwave images of the human body. As explained in Section 6.2, a fundamental problem in using EM fields for medical applications at microwave frequencies is that the microwave energy penetrates the body only slightly. This lack of penetration is a significant problem in microwave imaging. As a microwave signal travels from an external microwave applicator through the body to a microwave receiver on the other side of the body, it is so highly attenuated that little information can be obtained about the interior of the body. To get good resolution in the image, higher frequencies (shorter wavelengths) are better because the resolution is related to the wavelength, but as the frequency is increased, the EM fields are more and more attenuated by the body.

Microwave imaging has a significant opportunity in the detection of cancerous tumors and cardiac abnormalities because the electrical properties of normal and malignant tissue are significantly different, whereas their densities may be relatively similar (thus reducing the effectiveness of x-ray or MRI). In fact, there is some hope that the electrical properties change in precancerous regions and may be detectable prior to the tumor becoming well defined. Location of breast cancer shows particular promise, because the relatively low loss of breast tissue allows electromagnetic fields to propagate to the tumor and back, and the proximity of the tumor to the outer surface of the body means that the signal does not have more than a few centimeters to propagate.

Two major microwave imaging methods are available, and they are distinguished by how they map the region of interest. Tomography creates a complete voxel-by-voxel impedance image of the region, and confocal imaging maps only the significant scatterers. Impedance differences are needed for either method to distinguish organs, tissues, and so forth. Different tissues have different impedances, and tumors also have different electrical properties from normal tissue, thus enabling successful microwave imaging.

Tomography for Breast Cancer Detection

Microwave tomography provides a voxel-by-voxel image of the electrical properties in the region of interest. Since these properties vary from tissue to tissue, this image also generally corresponds to an image of the biological structures within the region of interest. Microwave tomography uses an array of antennas surrounding the region of interest to both transmit and receive the microwave signal. One such array is shown in Figure 3.58. One of the antennas in the array is used to transmit a signal, normally a sine wave, a set of sine waves, or a broadband signal (see Section 1.8), and all of the other antennas are used to receive the reflected and transmitted signals. Then another antenna transmits, and all of the others receive. And so on. Numerical dosimetry (see Chapter 5) is used to predict the data received from various slices through the model. These simulated data are then compared to the measured data, and the simulated model is adjusted until it gives a good match with the measured data. Several research teams have had good success with microwave tomographic imaging for breast cancer. In a system from Dartmouth, for instance, sine waves from 300 to 1,000 MHz (being expanded to 3 GHz) are transmitted from a circular array of sixteen transmit/receive monopole antennas, shown in Figure 3.58, to produce two-dimensional reconstructed images of the breast. Quarter-wave *(continued on next page)* monopole antennas, built by extending the inner conductor of semirigid coax, are used for this application, as described in Section 3.7. Water-filled waveguide apertures have also been used for tomography, but they are more difficult to construct.

Confocal Imaging for Breast Cancer Detection

Another microwave imaging technique that has shown excellent promise for breast cancer detection is confocal microwave imaging. Based on the theory behind ground-penetrating radar (GPR), confocal imaging seeks regions of large scattering rather than a map of electrical properties from tomography. This method can be applied in two ways. One is physically scanning (moving) a single antenna over the surface of the breast, recording signals in a grid pattern. The other (easier) method is to use a grid of many antennas placed above the breast. Each antenna in the array transmits in turn, and the signals are received by all other antennas in the array. These antennas are generally very small broadband antennas. For planar imaging, the patient lies face up, and the antenna array is placed in a plane above the breast. For cylindrical imaging, the patient lies face down, with the breast extending into a cylindrical array through a hole in the table. Matching fluid surrounding the breast is used to reduce scattering from the surface of the body. An ultrawideband pulse is transmitted by one antenna and received by the rest. Each antenna acts as the transmitter, in turn, and all of the others receive the scattered pulse that is reflected off impedance discontinuities (such as tumors, but also vessels and veins, and the skin-fat interface). The received signals are then time delayed depending on the distances between the transmitting antenna and the various receiving antennas. All of the received pulses that correspond to a specific location in space are then added up. The noise is reduced, and the signals from the scatterer add up, thus giving good visibility for small tumors. The more antennas and the higher the bandwidth, the better the image resolution can be. Therefore, the antennas must be small and work over an ultrawide frequency band. Imaging resolution of less than 1 cm requires a bandwidth of at least 5 GHz. High-frequency signals above 10 GHz do not propagate in the body, thus physically limiting the image resolution. Designing antennas for broadband applications such as confocal imaging is an exciting area of ongoing research. In general, the more varied paths for current that can be supported on a given antenna shape, the broader band it is. Bow-tie-shaped antennas, V's, crossed dipoles, and ridged horn antennas have all been proposed.

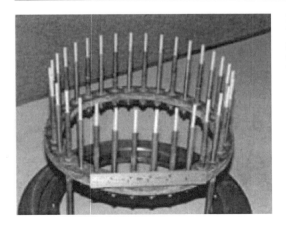

FIGURE 3.58

Two-dimensional monopole array used for tomographic imaging of the breast. (From Meaney, P. M., et al., *IEEE Trans. MTT*, 48, 1841–53, 2000. © 2000 IEEE. With permission.)

3.8 Diffraction

The simple planewave introduced earlier will propagate along a straight-line path perpendicular to its wavefront as it travels through a uniform medium (such as vacuum or a uniform dielectric) until it hits a reflecting surface or is bent at a refracting interface. This straight-line travel strictly holds, however, only for a planewave, which, according to the definition (see Section 3.2.2), has an infinitely wide wavefront. If the wavefront extent is limited—for example, by passing the planewave through a small slit or hole in an absorbing screen—the wave will start spreading outward after it passes through the aperture and will no longer be a planewave. This digression from straight-line travel (other than that caused by reflection or refraction) is termed *diffraction*.

3.8.1 Diffraction from Apertures

The extent of the spreading after passing through an aperture is inversely proportional to the size of the opening expressed in units of wavelength. Therefore, diffraction is not noticeable if the hole or transmitting aperture is many wavelengths wide. Diffraction becomes important only when the aperture is on the order of a few wavelengths or smaller. For the waves discussed in this chapter, the wavelengths range from about a millimeter to several meters, and typical openings may be comparable in size to these wavelengths. In this case, it is necessary to consider diffraction. Optical waves, on the other hand, as discussed in Chapter 4, possess wavelengths that are less than a micrometer, so diffraction occurs only for very small apertures. Diffraction also takes place when waves pass around an edge, wall, or some other single-sided boundary.

As an example of the effects of diffraction, consider the case of a planewave passing through a two-dimensional slit in an otherwise absorbing screen, as shown in Figure 3.59. The slit in this figure is ten wavelengths wide. The effect of the slit on the planewave can be thought of in this way: Each point in the aperture of the slit acts like a point source producing a spherical "wavelet." The transmitted beam is a combination of all those wavelets.

The transmitted beam's behavior can be divided into two regions. Just after passing through the aperture, the wave has not yet spread out appreciably, and the shadow of the slit still determines the approximate width of the beam. This region close to the slit is called the near field, or Fresnel region. One characteristic of this area is the irregular nulls and peaks in the magnitude of the **E** field (and associated **B** field, not shown). These spatial irregularities are caused by constructive and destructive interference between the wavelets propagating from all points in the aperture.

The behavior of the beam changes gradually as it travels away from the aperture going from the near field into the far field. The transition between these two regions is not abrupt, so the definition of this transition distance (already introduced in Equation 3.14) is somewhat arbitrary; however, it does provide a measure of the extent of the irregular, laterally confined near field.

The first change that is noticeable during the transition into the far field is that the beam field pattern becomes much more regular. The center portion of the beam (the main lobe) takes on a smooth, peaked shape, dropping to a null on both sides before rising again in repetitive side peaks (the side lobes) of rapidly diminishing size. The second change is that the edges of the beam—defined here by the first nulls on both sides of the main lobe as indicated by the dotted line—begin to spread laterally in a fashion that is linearly proportional to the distance from the aperture.

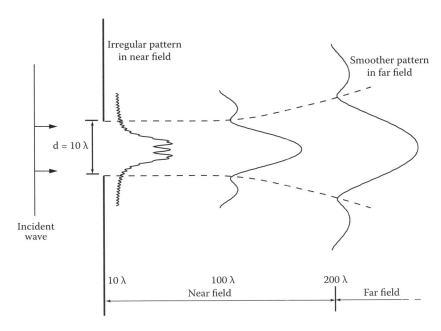

FIGURE 3.59
Diffraction effects on a beam passing through a slit aperture that is ten wavelengths (10 λ) wide. The profiles of the magnitude of **E** are plotted at distances of 10, 100, and 200 wavelengths from the slit. The **E** field
profiles show that the beam pattern is irregular in the near field, but becomes smoother in the far field. The
horizontal and vertical scales are not equal in this figure.

These characteristics (the transition distance and the degree of the linear spreading)
are determined by the size of the opening measured in units of wavelength. Figure 3.60
shows diffraction from an aperture that is half the width of that for the previous figure—
five wavelengths in this case. Note that the near field still shows irregularities, but over a
shorter extent than for the larger slit opening (as predicted by using D = d in Equation 3.14).
Also note that the spreading angle in the far field is almost double that of the previous
case. This angle, called the half angle of divergence, θ_d, is given for the two-dimensional
geometry shown in Figure 3.60 by

$$\sin(\theta_d) = \lambda/d \qquad (3.15)$$

where λ is the wavelength in the medium where the beam is propagating. Note that, importantly, the angle of spreading is inversely proportional to the size of the opening given as
a ratio to the wavelength. Thus, the smaller the opening or the longer the wavelength,
the larger the spreading. For three-dimensional diffraction (for example, from a circular
pinhole), the divergence angle and far-field beam pattern are not exactly the same as that
produced by a slit, but they show similar qualitative behavior.

The inverse relationship between divergence angle and opening size holds for other
situations as well, such as the beam propagating from a radio frequency power applicator
designed for hyperthermia treatment of cancer. At a frequency low enough to penetrate
deeply into the body to reach deep tumors, say 100 MHz, the wavelength in muscle is
about 30 cm. The aperture of a typical applicator will be on the order of this size, and
Equation 3.15 shows that the propagating beam will spread out considerably in the far

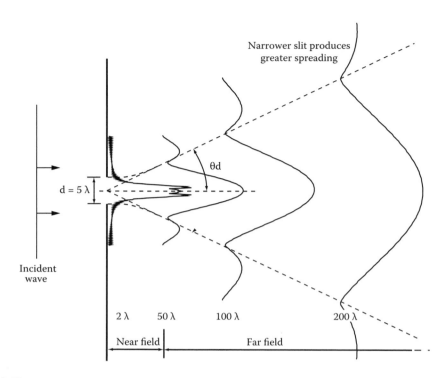

FIGURE 3.60
Diffraction effects on a beam passing through a slit that is one-half the width of Figure 3.59. The extent of the near field is smaller, but the angle of divergence of the **E** field profile is larger than before.

field, making localized energy deposition impossible at deep locations. Making the aperture smaller results in less radiation efficiency and in even more spreading in the far field, eventually approaching a uniform radiation pattern for a very small aperture like a point source. However, making the aperture smaller does confine the beam to a smaller dimension in the near field (though it is somewhat irregular).

3.8.2 Diffraction from Periodic Structures

An interesting wave interference effect takes place when there is an *array* of uniformly spaced apertures. This phenomenon is shown progressively in Figures 3.61 to 3.63. In Figure 3.61, the far-field pattern at an observation plane is shown for the radiation from two very narrow slits. (If there were just one narrow slit, the far-field pattern would be very wide, with nearly uniform illumination at the observation plane; the previous section has shown that the divergence angle is very large if the slit is narrow compared to the wavelength.) When two slits are present, there are some places on the observation plane where the two fields from the slits add in phase and the resulting **E** field is reinforced, and some places in between where the fields add out of phase (subtract) and are cancelled. The resulting pattern shows broad peaks with nulls in between.

When the number of slits is increased to four with the same center-to-center spacing, as in Figure 3.62, the constructive and destructive interference becomes more pronounced, causing the peaks to become narrower. Each peak is centered at the same position as

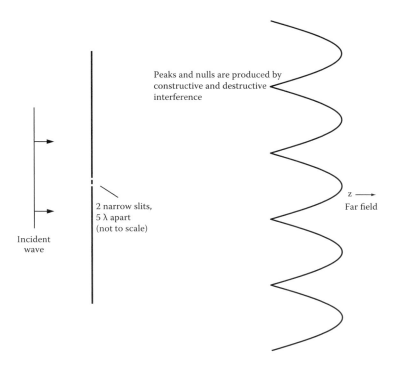

Peaks and nulls are produced by constructive and destructive interference

2 narrow slits,
5 λ apart
(not to scale)

Incident wave

z →
Far field

FIGURE 3.61
The magnitude of the **E** field in the far field after passing through two narrow slits. Constructive and destructive interference between the waves coming from each slit is responsible for the peaks and valleys.

before, because, as will be discussed shortly, the angles to the various peaks are set by the slit spacing in terms of wavelength, which has not changed. When the number of slits is increased to fifty, the peaks become very pronounced, as shown in Figure 3.63. Employing even more slits will make the peaks extremely narrow.

The angle of propagation of the peaks is an inverse function of the slit spacing (in units of wavelength). The derivation of this relationship is based on the following principle: the angles of maximum constructive interference (thus the angles of the peaks) are those angles where the paths taken by the waves from each neighboring slit differ by an integral number of wavelengths, and the waves therefore add in phase when they reach the observation plane. This situation is pictured in Figure 3.64 for two neighboring slits. When the path length difference, Δp, is equal to an integral number of wavelengths, nλ, trigonometry applied to the triangle gives the angle θ_n of the nth peak with respect to the horizontal axis as

$$\sin\theta_n = n\lambda/\ell \qquad n = 0, \pm 1, \pm 2..., \tag{3.16}$$

where ℓ is the spacing between the slits.

One of the most practical uses of this effect is the application of diffraction gratings in measuring the wavelength spectrum of an electromagnetic signal, most often in optics. In this case, the array of diffracting elements is usually composed of a grating of narrowly spaced grooves on a surface. When the beam to be analyzed is reflected from the grating, peaks corresponding to the various wavelength components in the beam are displayed on

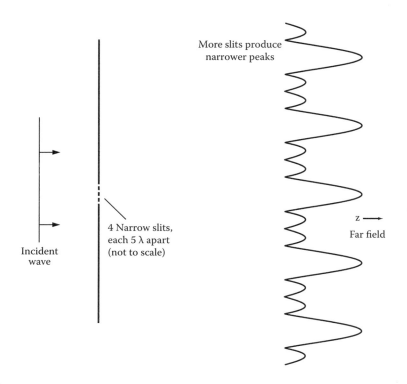

FIGURE 3.62

The magnitude of the **E** field in the far field after passing through four narrow slits with the same spacing as in Figure 3.61. The peaks are narrower, but are located at the same angles.

the detection plane at unique angles related to their wavelength by Equation 3.16. There are several spectroscopic uses of light in measuring the chemical or disease state of tissues (many *in vivo*), as discussed in the next chapter, which employ diffraction gratings for the analysis of the optical spectrum reflected or transmitted by the tissues.

3.9 Measurement of Mid-Frequency Electric and Magnetic Fields

At medium frequencies, the electric and magnetic fields are strongly coupled, which means that they mutually generate each other as they propagate. Thus, they are strongly related, and measurement of one is sufficient to understand both. In this frequency range, their biological effects are rarely considered independently.

As with low-frequency fields, mid-frequency measurements take an analog measurement and convert it to a digital value. Previous generations of measurement equipment utilized strictly analog values; however, this is very rare today. The mid-frequency electric field is typically measured with either antenna-like structures (as described in Section 1.2) or a waveguide or horn antenna. Mid-frequency magnetic fields can be measured with loops, although this is done far less often than for electric fields (because, as indicated above, measuring one field enables you to calculate the other in this band). The polarization of the electric field matters very much when receiving mid-frequency electric fields.

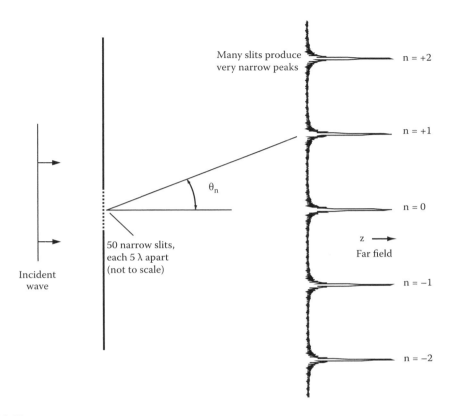

FIGURE 3.63
The magnitude of the E field in the far field after passing through fifty narrow slits with the same spacing as in Figure 3.61. The peaks become much narrower as interference from the multiple apertures sharpens the angular response. The peaks are labeled with an index n representing their spacing away from the central (n = 0) peak.

The receiving antenna must be oriented properly in order to receive these fields. For the case of electric field probes, three perpendicular antennas are usually used to receive the three separate components (x, y, z) of the electric field.

Mid-frequency electric fields are actually considered high frequency for electronics. (Standard voltage and current measurement equipment is unable to measure fields much above 1 to 10 MHz.) There are three major types of receiving equipment used in this band. The simplest is a *power meter* that measures rms power. Voltage or electric field (if desired) is proportional to the square root of the power. To measure the three components (x, y, z) of the electric (or magnetic) field, three separate power meters must be connected to each of the three antennas used to receive the field. A small and relatively inexpensive computer chip called a *received signal strength indicator* (RSSI) chip can also be used to measure rms power. These chips are used extensively in cellular telephone bands and are inexpensive in these bands because of the large volume produced. Both power meters and RSSI chips work over a relatively small frequency band. Power meters and RSSI chips tell the magnitude of power only, and nothing about the phase of the fields. They also generally measure all power in their band and cannot distinguish one frequency from another.

Spectrum analyzers are another way to measure the power in the mid-frequency band. A spectrum analyzer is connected to an antenna or waveguide to receive electric or magnetic

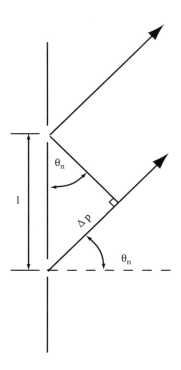

FIGURE 3.64
A simple picture of two neighboring slits that yields the angles θ_n of maximum constructive interference (i.e., the angles to the peaks). Maximum constructive interference takes place when the path length difference between the two rays, Δp, is an integral number of wavelengths.

fields and record the power as a function of frequency over a broad range. The spectrum of the power (how much power is at each frequency component) is displayed on a screen as well as stored digitally for output and further processing. Spectrum analyzers therefore distinguish one frequency from another (but do not usually measure the phase of each frequency component). Spectroscopy is an analysis of the power spectrum of a signal. Resonant frequencies display high power (indicating the presence of an element with that resonance), and nonresonant frequencies have lower power. Spectrum analyzers can be purchased for frequencies up to several tens of gigahertz today. They are much more expensive than power meters, because they can make more sophisticated measurements and usually have a wider frequency range. Virtually all spectrum analyzers have 50-ohm input impedance.

Vector impedance analyzers (or simply vector analyzers or network analyzers) are the most advanced (and expensive) measurement equipment used in this frequency band. Vector analyzers measure the magnitude and phase of the fields at all frequencies in their band. The vector analyzer is connected to a device or system being measured; it sends a transmitted sinusoidal electric signal down the cable to the device being tested, and receives the reflected signal coming back. The sinusoidal signal is scanned through all of the frequencies of interest, and the received signal collected at each frequency. The magnitude and phase of the reflected signal are measured in each case. Thus, the vector analyzer directly measures complex impedance, standing wave ratio, and magnitude and phase of the field over a broad range of frequencies.

Most vector analyzers have at least two ports, allowing them to do not just the reflection measurements described above, but transmission measurements where the signal is sent through a piece of hardware (such as into one end of a waveguide and out the other) as well. The signal that is transmitted to the second port is analyzed in the same way as the reflected signal for the first port.

The computer in the vector analyzer is able to digitally remove the mismatch between the cables and connectors and the device being tested. This requires calibrating the network analyzer: The cables and connectors that are to be used are connected to the network analyzer, and short, open, and 50-ohm (matched) loads are attached on the ends of the cables one at a time. Each measurement is stored in the network analyzer, which then uses network theory to determine the effect of the cables and connectors on the measurements. The network analyzer is also used with the dielectric measurement probe described in Section 3.3.2. The probe and cable are calibrated with a short circuit, an open circuit, and deionized water at the end of the probe. The computer in the network analyzer is then used to convert the reflected standing waves measured for the sample being tested into the permittivity and conductivity of the material at the end of the probe.

The network analyzer has a number of filtering and signal processing functions as well. One that can be quite useful is the ability to combine the frequency-domain measurements into a time-domain (pulsed) signature. This *time-domain reflectometry* signal is most often used to debug microwave circuits, but it has a number of other applications as well. The vector analyzer does its measurements in analog but converts them to digital for processing, storage, and output.

Oscilloscopes are also available in the mid-frequency range, although they are much more expensive than in the low-frequency range. They also convert the signals from analog to digital for output and processing. Oscilloscopes are used much less in the mid-frequency band than in the low-frequency band. Impedance in the mid-frequency band is virtually always measured with a vector analyzer, as impedance is almost always complex.

All mid-frequency electromagnetic hardware has a set frequency band in which it can be used, a minimum and maximum power level, and a certain input impedance. The input impedance is almost always 50 ohms. This means that if it is connected to a transmission line or other device that has a 50-ohm impedance, it will be matched and have no standing waves. But since most biological measurement systems are not 50 ohms, this mismatch results in a loss of power. All of the connections, connectors, and cables used in this frequency band are very critical and should be rated by the manufacturer for the frequency range of interest. Connectors and cables must be low loss at the frequencies being measured.

External signals are often measured inadvertently with the desired bioelectromagnetic signals. For this reason, most often coaxial cable is used. Coax has a shield around the inner conductor to reduce interference between the outside world and the signal on the cable. Unfortunately, the outside signal (such as a radio station, a cell phone, and noise from electrical systems turning on and off) is often still picked up by the antennas and leaks into the system from many places. Truly quiet electromagnetic measurements in this frequency range must be made inside an anechoic chamber. This is a metal room with foam on the walls that is designed to absorb electromagnetic fields.

Most measurement equipment in this frequency band has a very sensitive front end and can be easily damaged by sending excessive signal power into the equipment. Testing active devices (anything that has an amplifier, for instance) should be done with great care. Static electric discharge can also damage this equipment, so measurements are normally taken with the operator wearing a wristband connected to the equipment ground (outer metal shell) to prevent damaging the equipment.

COMPARING ULTRASOUND WAVES TO
ELECTROMAGNETIC WAVES IN THE BODY

Ultrasound waves are acoustic (sound) waves in a medium that transmit energy and information through vibrations of the medium's molecules and associated pressure. The prefix *ultra* denotes that these waves have a frequency higher than can be heard by humans, that is, higher than 20 kHz; typical frequencies used in biological applications are several hundred kilohertz to above 10 MHz. Ultrasound has found extensive use in medicine, especially for imaging of pregnancies, the heart, and abdominal organs. Pulse-echo imaging is the most common configuration, in which a pulse of ultrasound is launched into the body and the echoes that return from various organs and structures give an image of the internal anatomy, analogous to microwave radar. Moreover, ultrasound Doppler capability, which measures the frequency shift in reflections from moving scatterers such as red blood cells, allows blood flow in arteries and veins to be visualized and quantified.

What makes ultrasound imaging so attractive are its safety (at least at the power levels used for diagnostics), noninvasive nature, cost, and reasonable resolution. Recently high-power ultrasound energy has been proposed for a number of therapeutic applications, including drug delivery, thrombolysis, and tissue (e.g., cancer) ablation. Therapeutic ultrasound with MRI temperature monitoring is already finding clinical use for the treatment of uterine fibroids.

It is instructive to compare the characteristics of ultrasound as used in the body with those of electromagnetic (EM) energy; there are several similarities, but some important differences. First, the following are the things that ultrasonics and bioelectromagnetics have *in common*:

- Both can propagate as waves through the body. The ultrasound wave equation, in fact, is identical in form (but not in coefficients) to the EM wave equation, after some minor approximations are made in the basic ultrasound equations. A continuous ultrasound pressure wave follows the same form as given in Equation 1.9. This means that the concepts of frequency (Equation 1.10), phase velocity (Equation 1.13), propagation constant (Equation 1.14), and wavelength (Equation 1.15) apply to both modalities.
- There is a rough analogy between the pressure p associated with an ultrasound wave and the electric field E of an EM wave. Similarly, the particle (molecular) velocity u of an ultrasound wave is analogous to the magnetic field H of an EM wave. Thus, the concept of acoustic impedance ($Z = p/u$) is analogous to EM wave impedance ($Z = E/H$; see Section 3.2.1).
- Ultrasound waves undergo the same phenomena of reflection and bending at boundaries (Section 3.3.2) as EM waves, and can exhibit standing waves (Section 3.3.1) and resonance behavior. In fact, the reflections occurring at interfaces between tissues of different acoustic impedance are the basis of the ultrasound pulse-echo imager.
- Both ultrasound and EM waves attenuate as they propagate into tissue; most of the attenuation can be attributed to absorption (except for lung tissue, which has a considerable amount of scattering). Therefore, both modalities will lose power and consequently heat tissue as they *(continued on next page)*

propagate, to a degree dependent on frequency. The drop-off in wave amplitude (p for ultrasound, **E** for electromagnetics) has an exponential form versus propagation distance, as shown in Figure 1.35.

- The attenuation coefficient of both energy types increases (roughly linearly) with frequency. Defining effective penetration to be the depth at which the wave amplitude has dropped to $1/e = 37\%$ of its initial value at the body's surface, the frequency at which the penetration depth is equal to 10 cm is approximately 0.8 MHz for ultrasound, while for electromagnetics this frequency is about 70 MHz (see Figure 1.40).
- Both wave types can be focused or localized in the body to a size no smaller than their respective wavelengths, which by Equation 1.15 is inversely proportional to frequency. Therefore, there is a trade-off between penetration depth (the lower the frequency, the lower the attenuation and the deeper the penetration) and resolution (the higher the frequency, the smaller the wavelength and the better the resolution). See below for examples of wavelength values.

There also are some distinct *differences* between ultrasound and EM waves:

- The energy forms are completely different. Ultrasound is a manifestation of mechanical energy, involving pressure and particle (molecular) velocity as the main variables. Ultrasound needs a physical medium (usually water or soft tissue) through which to propagate and transmit energy in the form of vibrations of the medium's molecules. Electromagnetics, on the other hand, involves purely electrical and magnetic energy. It does not strictly need a physical medium for propagation; EM energy can travel through a vacuum (e.g., sunlight through outer space)—although it can, of course, propagate through the body, with some loss.
- Since the energy forms are different, the parameters entering calculations of wave properties are completely unique. This leads to a dramatic difference in the phase velocity for each wave type. For ultrasound, the speed of sound (in water or soft tissue) is 1.5×10^3 m/s, while the speed of light (in free space) is $c = 3 \times 10^8$ m/s, about 200,000 times faster. This in turn leads to large differences in wavelength. At the frequencies mentioned above corresponding to a 10 cm penetration depth, the wavelength of ultrasound is 2 mm, while the EM wavelength is 60 cm (after taking into consideration the relative permittivity of typical tissue at the specified frequency; see Appendix A and Equation 3.4). Thus, ultrasound has a much shorter wavelength and can be focused much more tightly; it fares better in the trade-off between penetration and focusing.
- The external applicators that introduce the waves into the body must be compatible with the respective energy types. Ultrasound uses electromechanical (piezoelectric) transducers, while radio frequency waves and microwaves typically employ loops, dipoles, or waveguide horns.
- The pressure p of an ultrasound wave is a scalar quantity, while EM field quantities (**E** and **H**) are vectors with three components each at any point in space.

3.10 Summary

In the mid-frequency range described in this chapter, the electromagnetic fields act like waves rather than voltages and currents (Chapter 2) or rays (Chapter 4). They are strongly coupled, and interact with the body in very complex ways. This chapter has shown the basic mechanisms with which waves interact: reflection, transmission, attenuation, diffraction, and constructive and destructive interference. Each of these effects is straightforward to understand, but in typical applications their combinations yield field patterns sufficiently complex that detailed numerical simulation codes are typically used; these numerical methods are discussed further in Chapter 5.

The applications in this area are very diverse and include many different methods for producing heat, communication, and imaging. Controlling the fields in this frequency band can also be challenging, because they tend to "leak." Thus, methods to provide shielding and grounding are specialized in this frequency region. Applications in the mid-frequency region continue to expand and be refined, as will be discussed further in Chapter 6. The next chapter will discuss higher-frequency fields, where the fields behave more like rays than waves.

4

EM Behavior When the Wavelength Is Much Smaller Than the Object

4.1 Introduction

In this chapter, we discuss the case where the wavelength of the electromagnetic radiation is much smaller than the size of typical objects. Since details smaller than about 0.3 mm are difficult to resolve with the naked eye, this chapter is concerned with waves whose frequencies are high enough that their wavelengths are smaller than 0.3 mm. Since $f = c/\lambda$ (Equation 1.15), this means that the frequency will be in the range of 3×10^8 (m/s)/0.3×10^{-3} (m) $= 1 \times 10^{12}$ (Hz) = 1 THz and higher. The lowest end of this frequency range intersects with the millimeter-wave band, so named because its wavelengths are fractions of a millimeter up to a few millimeters (see Figure 1.31 for a graph of the various electromagnetic regions). At somewhat higher frequencies are the far-infrared waves, then the near-infrared waves whose wavelengths are on the order of micrometers, named far and near according to their relative closeness to visible light wavelengths. Higher in frequency (shorter in wavelength) is the very important visible wavelength range (between 400 and 700 nm), where many significant discoveries and devices such as lasers have been made, undoubtedly due to the significance of light in human vision. At slightly higher frequencies are the ultraviolet (UV) waves. At much higher frequencies are the soft, then hard, x-rays.

All of these waves share some common characteristics, but since the frequency range covered is so broad, there are distinct differences as well. For example, the millimeter waves at the low end of this range are often considered to be at the high end of microwaves (Chapter 3) since they use generators, detectors, and waveguides that are specialized versions of the corresponding microwave devices. On the high end of this range, x-rays possess such high frequencies and short wavelengths (thus their extensive use for medical imaging) that they are not appreciably refracted or slowed down by materials containing dipoles (such as water-based tissues); rather, they are absorbed and scattered by individual atoms. This absorption has the capability to transfer enough energy to the atoms to ionize them. Thus, x-rays are called *ionizing radiation*, and they behave much differently from lower-frequency electromagnetic waves. Lower-frequency waves are nonionizing, and do not damage cells through ionization effects.

In between these extremes are the *optical* waves, which by most definitions encompass the infrared, visible, and ultraviolet regions. These waves are the main focus of this chapter. They share many common features, and play an important role in human life. Since their wavelengths (from one-tenth to a few micrometers) are much smaller than the typical object, it is often convenient to describe them in two ways that are different from those used in the previous chapters. The first difference is that diffraction is often unnoticeable for these waves (unless purposely caused by specialized diffraction gratings or other very small structures). This is because, as covered in Section 3.8, the degree of diffraction is proportional to the ratio of wavelength to object size. Since the wavelength is short compared to most objects (for example, a mirror), diffraction is small and does not play a major role in the behavior of the propagating wave. The waves then can be conveniently described by straight-line *ray* propagation, and *ray tracing* is used to make optical propagation much easier to visualize and determine. This is the domain of *geometrical optics*, where the rays follow geometrical rules. Ray tracing is used extensively in the first half of this chapter.

The second difference is that instead of using the permittivity coefficient ε to describe the effect of a material on the propagation of an optical wave, the square root of the relative permittivity ε_r is used. This term is defined as the *index of refraction* n:

$$n = \sqrt{\varepsilon_r} \tag{4.1}$$

The usefulness of this definition stems from the fact that refraction, as given by Snell's law (Equation 3.5), plays an important role in optics, where lenses are used to focus or change the wavefront curvature of the waves. Snell's law can be conveniently restated in terms of the index of refraction n. In Figure 4.1, let the wave incident on an interface between two different dielectric materials be represented by its ray at an angle of incidence θ_i (by convention measured with respect to a line perpendicular, or normal, to the interface). Let the relative permittivity of material 1 be ε_{r1} and the relative permittivity of material 2 be ε_{r2}. After passing through the interface, the angle of transmission is θ_t. Since $\varepsilon_{r1} = \varepsilon_1/\varepsilon_0$ and $\varepsilon_{r2} = \varepsilon_2/\varepsilon_0$ (see Section 1.6), Equation 3.5 can be formulated in terms of relative permittivity as

$$\sqrt{\varepsilon_{r1}} \, \sin \theta_i = \sqrt{\varepsilon_{r2}} \, \sin \theta_t. \tag{4.2}$$

Then, using the definition of index of refraction, Equation 4.1, Snell's law becomes simply

$$n_1 \sin \theta_i = n_2 \sin \theta_t. \tag{4.3}$$

This is the form of the equation usually seen in optical analyses.

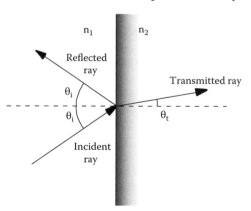

FIGURE 4.1
The refraction (bending) of a ray transmitted through an interface between two dielectrics with refractive indices n_1 and n_2, respectively. In this example, $n_2 > n_1$.

Also note that, using Equation 3.4, the phase velocity of the wave in each medium is given by

$$v_{p1} = c/n_1 \text{ and } v_{p2} = c/n_2 \qquad (4.4)$$

Thus, the index of refraction is a measure of how much slower a wave is in a medium compared to free space. It follows that Equation 4.3 can equivalently be written as

$$v_{p2} \sin \theta_i = v_{p1} \sin \theta_t. \qquad (4.5)$$

All values of refractive index must be greater than 1, since the law of special relativity states that no energy wave can travel at speeds greater than the speed of light in free space, c. Generally, the denser the medium, the higher the index of refraction and the slower the wave speed. Typical values of n are 1.33 for water, approximately 1.5 for glass (depending upon the type of glass and the wavelength of the light), and approximately 1.4 to 1.6 for various transparent plastics. For tissues that have high water content, n is somewhat higher than 1.33, depending upon the density of the proteins, fibers, and other constituents of the tissue. When a substance is lossy, its refractive index becomes a complex number.

4.2 Ray Propagation Effects

As a consequence of the lack of diffraction for most optical waves, as explained in the previous section, the waves will travel in a straight line until reflected (at a mirror, for example) or refracted (by a lens). Due to the smallness of the wavelength, the waves behave like segments of a planewave over any extent of practical interest. Earlier, in Figure 3.3, it was seen that the propagation vector **k** is perpendicular to the wavefront of a propagating planewave. The direction of the propagation vector **k** defines the direction of travel of the wave; this vector direction defines the *ray* associated with that wave. Thus, when diffraction can be neglected, the ray in a uniform medium travels in a straight line until being redirected.

4.2.1 Refraction at Dielectric Interfaces

It is very useful to use this straight-line ray behavior to describe how optical waves are refracted and focused. For example, Figure 4.1 has already used ray tracing to show how an incident ray is refracted into a different angle after passing through a dielectric interface. For this particular figure, it was assumed that the index of refraction n_1 was smaller than the index n_2, as when going from air ($n_1 = 1$) into water ($n_2 = 1.33$). In this case, the ray was bent more *toward* the normal (perpendicular) direction. As predicted by Equation 4.3, when $n_2 > n_1$, $\sin \theta_i > \sin \theta_t$ and $\theta_i > \theta_t$.

Figure 4.2 shows the opposite case, where n_1 is larger than n_2, as when going from glass ($n_1 = 1.5$) into air ($n_2 = 1$). In this case, the ray is bent more *away* from the normal after transmission.

A MEMORY AID FOR RAY BENDING

It is often very useful in doing ray tracing to be able to quickly remember which direction a ray will tilt when transmitted in each of these cases. Figure 4.3 gives a very simple, qualitative memory aid for determining the direction of refraction. Think of the ray as being an army tank with two independent treads, one rotating on each side. The tank is traveling in the direction of the ray. Remember that the speed of the wave (the tank and its treads) is inversely proportional to the index of refraction of the material, as given by Equation 4.4. Thus, if $n_2 > n_1$ as in Figure 4.1, the speed is slower in medium 2 than in medium 1. Now, as the tank crosses the boundary into medium 2, the right tread will enter first and therefore will be slowed down before the left tread is slowed. This will cause the tank to veer more toward the normal, consistent with Snell's law. If $n_1 > n_2$ as in Figure 4.2, medium 2 is faster than medium 1 and the right tread will speed up before the left tread. The tank will veer farther away from normal, again consistent with Snell's law.

Even though this memory aid is very simple, it actually can be used to obtain Snell's law quantitatively. It also predicts correctly the event when a wave is incident exactly perpendicular to any dielectric interface, as shown in Figure 4.4. In this case, both treads enter at the same time and the tank is not turned either way. In other words, when a wave is incident normally (i.e., perpendicular to an interface), the wave continues on with no bending. This is true regardless of the relative magnitudes of the indices of refraction. Of course, Snell's law also predicts this behavior, since when $\theta_i = 0°$, then $\sin \theta_i = 0$, and thus $\sin \theta_t = 0$ and $\theta_t = 0°$, independent of the values of either n_1 or n_2. Incidentally, Snell's law also shows that refraction of the incident ray will happen only when n_1 is different from n_2; if the two regions have the same index, there will be no refraction of the ray, as may be intuitively obvious.

Returning to the case when $n_1 > n_2$, Figure 4.2 can be used to predict the onset of total internal reflection (TIR), a phenomenon that has major importance in fiber optics. TIR has been discussed in Section 3.3.2, and here we reformulate it using refractive index terminology. TIR occurs when the incident angle is large enough that the transmitted angle is 90° and essentially the transmitted beam disappears. The incident angle at which this

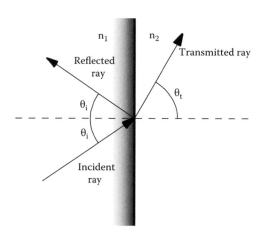

FIGURE 4.2
The refraction (bending) of a ray transmitted through an interface between two dielectrics with refractive indices n_1 and n_2, respectively. In this example, $n_1 > n_2$.

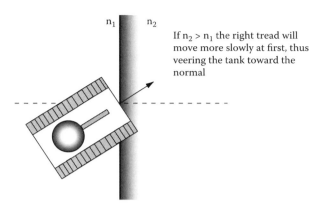

n_1 n_2

If $n_2 > n_1$ the right tread will move more slowly at first, thus veering the tank toward the normal

FIGURE 4.3
A simple memory aid to determine which direction a ray will bend when striking a dielectric interface. The army tank represents the incoming wave. One tread will enter region 2 before the other tread, and will be either speeded up or slowed down, depending *inversely* on the relative values of n_1 and n_2. A larger index n means a slower velocity; a smaller index n means a faster velocity. This causes the tank to veer to one direction or the other.

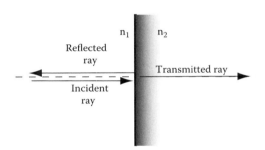

n_1 n_2

Reflected ray

Transmitted ray

Incident ray

FIGURE 4.4
The situation when a ray strikes the interface at normal incidence (perpendicular to the boundary). Both the transmitted ray and reflected ray are also perpendicular to the boundary, and there is no refraction.

happens is called the *critical angle*, and can be found from Equation 4.3 by setting $\theta_t = 90°$, and therefore $\sin \theta_t = 1$. Then,

$$\sin \theta_{ic} = n_2 / n_1 \qquad (4.6)$$

TIR will take place whenever the incident angle is equal to the critical angle given by Equation 4.6 or larger, including up to an incident angle of 90° (grazing). Over this entire range, there is no transmitted ray into medium 2, only a totally reflected ray in medium 1. Incidentally, since $\sin \theta_{ic}$ has to be smaller than unity for any realizable incident ray, Equation 4.6 shows that TIR will occur only when $n_1 > n_2$.

4.2.2 Optical Polarization and Reflection from Dielectric Interfaces

The orientation of the incident **E** field vector defines the *polarization* of the incident wave. As shown earlier in Figure 3.3, the **E** field vector will be perpendicular to the direction of propagation of the wave (in most cases of interest in optics), and therefore to the ray. But the orientation of **E** still needs to be specified in the plane perpendicular to the ray. There are two distinct possibilities for the orientation of **E** when dealing with reflection, as shown in Figure 4.5. The plane used for referencing the orientation of **E** is called the *plane*

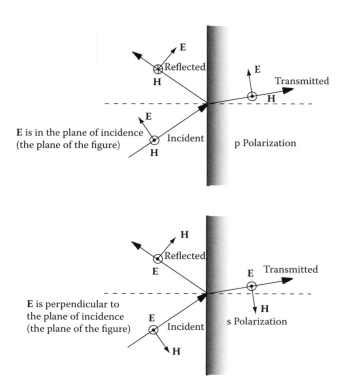

FIGURE 4.5
Definitions of the two independent states of polarization of the incident **E** field. For p polarization, the **E** field vector lies in the plane of incidence (the plane of the figure). For s polarization, the **E** field vector is perpendicular to the plane of incidence.

of incidence, and is defined as the plane that contains both the incident ray and a line drawn perpendicular to the interface (i.e., the dashed line). In Figure 4.5, the plane of incidence is the plane of the paper. It also contains the reflected ray and the transmitted ray.

When the **E** field vector is in the plane of incidence, this is *p polarization*. When the **E** field vector is perpendicular to the plane of incidence (pointing either in or out), this is *s polarization*. Generally, any arbitrary incident polarization can be written as a combination of these two states.

You may have noticed in both Figures 4.1 and 4.2 that in addition to the transmitted ray there is also a reflected ray. This is true for all angles of incidence, not just for TIR. As discussed in Section 3.3.2 for microwaves, the angle of reflection is equal to the angle of incidence in order to match boundary conditions at all locations along the interface. The amplitude of the reflected wave is found from the *Fresnel* formulae; reflection from a dielectric interface is called *Fresnel reflection*. The magnitude of the reflection is specified by the amplitude reflection coefficient ρ, which is the ratio of the amplitude of the reflected **E** field to the amplitude of the incident **E** field. ρ depends strongly on the two refractive indices, the angle of incidence and the polarization of the incident wave. For p-polarized light, the amplitude reflection coefficient is

$$\rho = \frac{\tan(\theta_i - \theta_t)}{\tan(\theta_i + \theta_t)} = \frac{\left|E_{\text{reflected}}\right|}{\left|E_{\text{incident}}\right|} \tag{4.7}$$

where the value of θ_t can be obtained from Snell's law. In the limit of normal incidence (i.e., for $\theta_i = 0°$), Equation 4.7 reduces to

$$\rho = \frac{n_2 - n_1}{n_2 + n_1} \tag{4.8}$$

For s-polarized light, the amplitude reflection coefficient is

$$\rho = \frac{-\sin(\theta_i - \theta_t)}{\sin(\theta_i + \theta_t)} = \frac{|E_{reflected}|}{|E_{incident}|} \tag{4.9}$$

In the limit of normal incidence ($\theta_i = 0°$), Equation 4.9 reduces to

$$\rho = \frac{n_1 - n_2}{n_1 + n_2} \tag{4.10}$$

When ρ has a negative value, the amplitude of the reflected wave is 180° out of phase from the incident wave.

These reflection coefficients (ρ) are for the E field. To get the *power* reflection coefficient R, or *reflectivity* (the ratio of reflected power density to incident power density), ρ must be squared, since power density is proportional to E^2, as explained in Section 3.4.3. Thus, $R = \rho^2$.

As predicted by Equations 4.7 through 4.10, there is some amount of reflection whenever an optical wave passes from one dielectric medium into another. That is why you can see a faint reflection of yourself whenever you look through a glass window. There are actually two separate reflections, one from the front air-glass interface and one from the back glass-air interface. Assuming normal incidence and putting typical values for the indices of air ($n_1 = 1$) and glass ($n_2 = 1.5$) into Equation 4.8 gives $\rho = 0.2$, so $R = \rho^2 = 0.04$. Thus, 4% of the incident power is reflected from the front surface of the glass. Interchanging the roles of n_1 and n_2 changes the sign of ρ, but because $R = \rho^2$ there is no difference in the magnitude of R, so 4% is also reflected from the back surface and 8% of the power is reflected by both surfaces combined. In some situations, such as eyeglasses, special antireflection coatings can be deposited on the glass to reduce this amount to nearly zero.

There are two special cases of reflection that can be determined (with some effort) from Equations 4.7 and 4.9. First, for both polarization states, $\rho \rightarrow 1$ when $\sin\theta_i = n_2/n_1$. This is just a restatement of the condition for TIR discussed earlier. Since $\rho = 1$, it shows that reflection really is *total* (i.e., it is 100%—not merely 99% or 99.99%), as the name *total internal reflection* implies.

The second special case occurs only for p polarization. *Brewster's angle* θ_B is defined as that angle of incidence at which

$$\tan\theta_B = n_2/n_1 \tag{4.11}$$

At this special angle of incidence, the magnitude of the reflected wave given by Equation 4.7 goes identically to zero; there is no reflected power. As an example, Figure 4.6 plots the variation of R as a function of the angle of incidence for both polarization states for $n_1 = 1.5$ and $n_2 = 1$. The drop in reflectivity to zero at Brewster's angle can clearly be seen for the

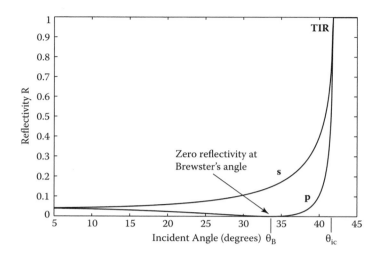

FIGURE 4.6
The variation of the power reflectivity R from a dielectric interface as a function of the angle of incidence. In this example, $n_1 = 1.5$ (glass) and $n_2 = 1$ (air). Note that p-polarized light has zero reflectivity at Brewster's angle ($33.7°$ for this example). Also, note the rise to TIR at the critical angle ($41.8°$ for this example).

p-polarized light, as can the rise to unity at the critical angle for both polarization states for TIR.

4.2.3 Ray Tracing with Mirrors and Lenses

Optical mirrors are usually made of conductive metal coatings (such as aluminum) on a substrate such as glass. The reflection of optical waves from conductive coatings follows the same boundary conditions as for lower-frequency microwaves (see Section 3.3.1). The reflectivity R depends upon the conductivity of the coating, but is generally around 90% for aluminum, and higher for silver and gold. As before, the angle of reflection is equal to the angle of incidence. This is known as *specular* reflection. It makes ray tracing of the paths of the reflected waves from mirrors easily visualized. For example, Figure 4.7 shows parallel incident rays reflected from both a flat mirror and a spherical mirror. Since the spherical mirror surface is curved inward (concave), specular reflection causes the parallel rays to focus to a point halfway between the center of curvature and the mirror surface. (This is true for rays near the axis; for rays farther toward the periphery, some spreading of the focus spot, known as *aberration*, occurs.) Thus, for a concave spherical mirror, the focal length f is equal to one-half the radius of curvature R_0, or $f = R_0/2$.

Lenses are most often used to focus or expand light beams. The direction and degree of curvature of the lens surface determines, following Snell's law (Equation 4.3), the amount of focusing of the lens and whether it will converge or diverge an incoming beam. Several varieties of lenses are shown in Figure 4.8. If a lens causes an incoming parallel set of rays (called a *collimated* beam) to converge, it has a positive focal length and is called a *positive* lens. If the lens causes the collimated beam to diverge, it has a negative focal length and is a *negative* lens.

The labeling of lenses usually obeys the following convention: the type of the curvature of the first surface of the lens as viewed from one side (concave, convex, or plano) is given first, then the curvature of the other lens surface as viewed from the other side is

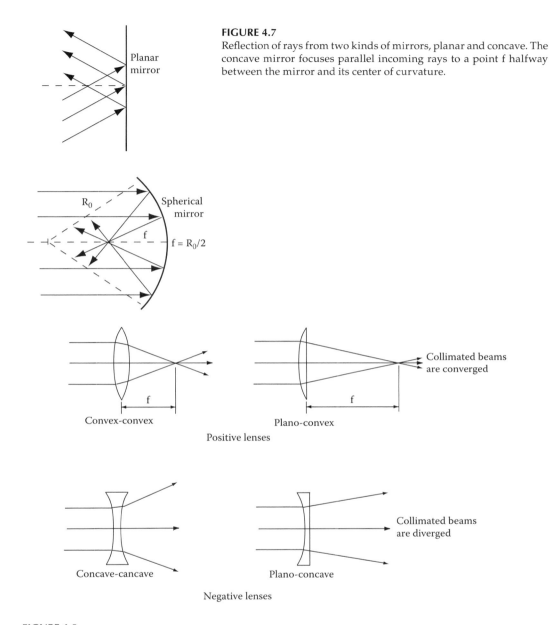

FIGURE 4.7
Reflection of rays from two kinds of mirrors, planar and concave. The concave mirror focuses parallel incoming rays to a point f halfway between the mirror and its center of curvature.

FIGURE 4.8
Examples of both positive and negative focal length lenses; f is the focal length of the lens. The convention for naming the lenses is given in the text.

given next. Thus, a convex-convex lens (sometimes called a double-convex lens) has two surfaces that bulge out at the center, as in the upper left of Figure 4.8. Using Snell's law (or qualitatively, the tank memory aid), you can easily trace a ray through a convex-convex lens to see that it will refract a ray originally parallel to the axis to a focal point on the axis on the other side of the lens, and therefore it is a positive lens.

For thin lenses, it does not matter significantly which orientation a lens takes with respect to the propagation direction of the rays. In other words, if you turn a plano-convex lens around to make it a convex-plano, it will still focus the incoming rays to the same spot. However, aberrations are reduced if the collimated beam strikes the convex surface first, then passes through the planar surface second; that is, it passes in the direction shown in the upper right of Figure 4.8. This keeps the maximum refraction angle encountered by any ray to a smaller value than if the lens were the other way around.

The focusing power of a lens is determined by both the degree of curvature and the index of refraction of the lens material. Thus, a lens with steeper curvature (or two curved surfaces instead of just one) will focus more strongly; this is indicated by a shorter focal length f. Also, a lens with a higher refractive index will focus more strongly than one with lower index. This is why lightweight plastic eyeglass lenses are sometimes made of polycarbonate material; the high refractive index of polycarbonate (n = 1.6) allows the lenses to be curved less (with corresponding lower overall thickness and weight) than with glass or lower index plastics.

A common way to specify the focusing power (or equivalently, the light gathering capability) of a lens is by its f-number, sometimes given as f# or f/. The f-number of a lens is defined by

$$f\# = f/D \tag{4.12}$$

where f is the focal length of the lens and D is the diameter of the lens or the diameter of a limiting pupil, whichever is smaller. The smaller the f-number, the bigger the cone of focused rays, or when used to collect light from a broad-angle source, the more light the lens will accept.

4.2.4 Imaging with Lenses

The major use of lenses in optics is to form images of objects, as in telescopes or cameras. Figure 4.9 shows a simple imaging configuration in which the image of an arrow is formed by a positive lens (in this case, a plano-convex lens). There is some magnification M of the image size ℓ_i compared to the original object size ℓ_o, and this magnification is defined as $M = \ell_i/\ell_o$. For a given imaging setup, the amount of magnification is determined by the ratio of the image distance d_i to the object distant d_o, or $M = d_i/d_o$, as can be seen by using similar triangles in Figure 4.9.

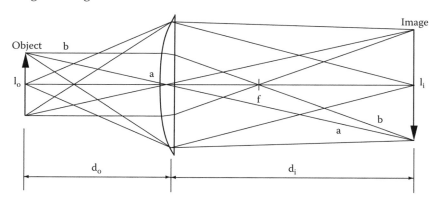

FIGURE 4.9
A simple imaging arrangement using a thin plano-convex lens. The image of the object is magnified by an amount M. The intersection of the rays labeled a and b determines the location of the image, as explained in the text.

You can quickly get the position of the image in Figure 4.9 by the following simple ray tracing procedure: Trace one ray from some point on the object (say the arrow tip, the ray labeled *a*) through the center of the lens; it will not be bent overall (for a thin lens) since it passes directly through the lens center. Then trace another ray from the same point (ray *b*) that goes parallel to the axis until it passes through the lens; since it is initially parallel to the axis, the lens will redirect it through the lens focal point, at a distance f from the lens. Where these two rays cross on the other side of the lens, the image of the chosen object point will be formed. Other rays from the same object point will also meet at this image point (in the absence of aberrations). Then d_i can be measured and M determined.

The *lens law* can also be used to solve for the image distance d_i. This law relates the three distances important in image formation for a thin lens:

$$\frac{1}{d_i} + \frac{1}{d_o} = \frac{1}{f} \qquad (4.13)$$

Specifying two of the three distances will allow you to solve for the third.

SUNGLASSES

Sunglasses are used in everyday life to protect our eyes from the sun and glare, and they combine many of the optical properties in this chapter. Figure 4.10 shows the several different layers of coatings that make up a good pair of sunglasses. Let us examine each layer and see how it uses the principles of optics:

Mirroring: The outermost layer of sunglasses is often a thin-film metallic coating. Although the outer layer may look like a perfect mirror, it is not (otherwise, no light would reach your eye). The mirror is convex, which makes people on the outside see their reflection as very distorted. (The closer you get to the convex mirror, the larger your nose appears and the smaller your feet.) Sometimes this reflective layer is uniform, but other times it is graded. In most environments, more light comes from above (the sky) than below. For water skiing and snow skiing, however, a great deal may also reflect up from the water or snow below. Thus, many sunglasses have more reflectivity at the top and bottom of the lens to reduce light coming from above and below, and less through the center, where presumably you are trying to see. Some sunglasses are even made with a holographic type design fashioned into the mirroring, so that from the outside the sunglasses appear to have odd patterns. Looking from the inside, the design is not seen.

Antiscratch coating: The mirrored coating on sunglasses is backed by an antiscratch coating to protect the rest of the sunglass lens.

Polarizing film: Polarization tells the direction of the electric field vector. Normal sunlight is randomly polarized, which means it contains all polarization directions about equally. But when light reflects off of water, snow, or other highly reflective objects, the part of the light that is polarized parallel to the ground (s polarization) is more strongly reflected than the portion polarized perpendicular to the ground (p polarization). Thus the reflected glare is mostly s-polarized, and can be eliminated with a polarizing filter.

(continued on next page)

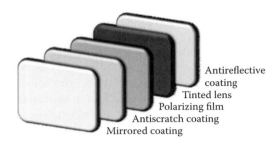

FIGURE 4.10
Please see color insert following page 146. Layers commonly found on sunglass lenses.

Antireflective
coating
Tinted lens
Polarizing film
Antiscratch coating
Mirrored coating

A polarizing filter is made by inserting iodine into a long-chain polymer film, then aligning the polymer's long chains by stretching. Light that is polarized parallel to the polymer's long axis is at least partially blocked, but light with polarization perpendicular to the long axis passes through. The reason is as follows: Electrons from the iodine impart high local conductivity in a preferential direction along the polymer's long axis. An electric field parallel to this axis is then tangential to the direction of high conductivity and, by boundary conditions, must go to zero (see Section 3.3.1). This causes the parallel polarization component to suffer high reflection losses, but not the perpendicular polarization component.

This polarizing film reduces much of the glare from reflections off of the ground. Removing the glare reduces the intensity of the light (which is nice for skiing, for example), and it allows you to see beyond the reflecting surface (to see fish and other objects under the water, for instance). Polarizing filters are also used on cameras to increase the intensity of the blue in the sky.

Tinted lenses: Tinting of the lenses is used to reduce the overall brightness that reaches your eye. Gray tints are the most common, because they reduce the overall brightness with minimal distortion in other colors. Amber or brown lenses also reduce brightness, and they absorb blue and UV light, which have been shown to cause cataracts. Yellow lenses absorb almost all of the blue light and make objects appear bright and sharp, although the colors are significantly distorted. Most ski goggles use yellow or gold lenses. Green lenses also absorb blue, and are known for having the highest color contrast. Purple and rose lenses give emphasis to objects outdoors against green and blue backgrounds. The phrase "Looking at the world through rose-colored glasses" refers to the apparent enhancement of beauty these lenses provide. In addition to tinting, it is important to have sunglasses with a coating to filter 100% of the UV rays to prevent cataracts and other disorders of the eye.

If the sunglasses are nonprescription, the lens itself has zero *net* curvature, meaning that the two surfaces of the lens are curved by the same amount, imparting no focusing power to the lens. But the lens can also be fabricated with optical *power* (focusing, Figure 4.8) to correct the vision of the wearer.

Antireflective coating: This coating is comprised of a number of half-wave dielectric layers deposited on the inside of the lens. Each half-wavelength layer prevents reflection at one frequency, as shown in Section 3.3.2 and Figure 3.20(c). Several layers back-to-back, each at a slightly different thickness, act as a complete-transmission, no-reflection coating. Their purpose is to reduce the annoying reflection of your eyes off of the inside surface of the glasses.

4.2.5 Graded-Index Lenses

There is a circumstance in which a ray will not follow straight-line propagation, even though diffraction can be neglected. This is the case when the ray is traveling in a medium whose refractive index is not uniform, but rather varies spatially. This type of medium has a graded refractive index, and the ray will follow a curved path whose direction and degree of curvature depend upon the index profile.

One of the practical uses of this concept is in the *graded-index lens*, or GRIN lens. Here the focusing power of the lens is not provided by curved faces, but by the graded variation of the lens's refractive index. Usually made from doped glass, the lens has a high index at the center and the index decreases radially toward the edges. A ray will then take a path similar to that shown in Figure 4.11. The ray will follow the path that gives the shortest propagation time between two points (say A and B in Figure 4.11); this is known as Fermat's principle. Since the velocity of the ray is inversely proportional to the index of refraction, as explained earlier, the propagation time is actually shortened by a path that takes the ray out into the lower-index margins rather than going straight through the high-index center. As seen in the figure, rays coming from a point on one side of the lens will be refocused to a point on the other side of the lens. The GRIN lens thus acts as a true lens and can be used in situations where a regular lens is not suitable.

The GRIN lens is usually much smaller than a regular spherical lens, and is often used at the ends of optical fibers (next section) and employed in endoscopes or in other applications where the light from a fiber must be focused, as in fibers for laser light delivery to tissues. Also, GRIN lenses can be fabricated to have less aberration than regular spherical lenses.

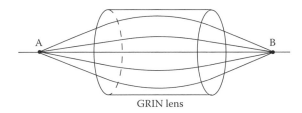

GRIN lens

FIGURE 4.11
A graded-index lens, or GRIN lens, which refracts rays by index variation rather than surface curvature.

4.3 Total Internal Reflection and Fiber Optic Waveguides

One of the most important and practical uses of total internal reflection (TIR) is in optical waveguides. When these waveguides are round glass or plastic fibers, they are known as optical fibers or fiber optic waveguides. We will use glass fibers as the main example of optical waveguides, although other geometrical forms, such as planar waveguides, are also sometimes seen.

The side view of a round optical fiber is shown in Figure 4.12. The central region of the fiber, known as the *core*, is composed of glass with refractive index n_1. The diameter of the core depends upon whether the fiber is intended to be multimode or single mode (discussed later); multimode fibers have diameters that generally range from 50 to 200 μm. Formed around the core is a layer of lower-refractive-index glass, called the *clad*,

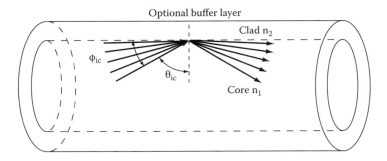

FIGURE 4.12
The structure of a multimode optical fiber. The core refractive index n_1 must be greater than the clad index n_2 in order to have TIR between the core and the clad. All rays traveling inside at angles that are less than the critical angle ϕ_{ic} with respect to the boundary will be trapped, as shown by the cone of rays.

with refractive index n_2. This forms a step-index fiber. It is essential that $n_1 > n_2$ for TIR to occur. When this is the case, rays inside the core will be trapped by TIR if they strike the core-clad interface at shallow enough angles. From Section 4.2.1, remember that TIR will occur whenever a ray is incident on a dielectric interface with an angle (measured with respect to the perpendicular) that is at or exceeds the critical angle given by Equation 4.6.

It is often convenient in waveguides to define angles with respect to the boundary plane (interface plane), denoted by ϕ, rather than with respect to the perpendicular line, denoted by θ. Thus, ϕ and θ are complementary angles and add to $90°$. Rewriting the critical angle equation in terms of this angle gives

$$\cos \phi_{ic} = n_2 / n_1 \qquad (4.14)$$

Then all rays that are propagating inside the core at angles that are at or less than ϕ_{ic} with respect to the boundary will be trapped by TIR and will remain inside the core until they reach the end of the fiber and radiate out the end face. This is indicated in Figure 4.12 by a cone of trapped light rays. The attenuation of light by modern glass fibers (due to absorption and scattering) is very low, around 3 dB/km for visible light and less than 0.2 dB/km for certain infrared wavelengths, so very little light is lost. As emphasized in an earlier section, TIR means total reflection, so no light is lost by transmission out the sides of the fiber (unless the fiber is tightly bent; then there may be some so-called bending losses).

The need for the clad layer may be puzzling, since if the core is situated in air (thus $n_2 = 1$), the condition that $n_1 > n_2$ for TIR would surely be met. This is quite true if it could be assured that air would always surround the fiber. But in practical situations, the core will touch other objects, such as metal, plastic, or even fluids. At these locations, n_2 may then be greater than n_1, and TIR would be frustrated and the previously trapped rays would be lost out the side of the fiber. The presence of the clad avoids this possibility by protecting the core-clad interface.

How thick should the cladding be? Although the simple ray picture does not predict it, not all the **E** field of the trapped rays is actually confined to the core. A small portion of the field, called the *evanescent tail*, extends a few hundred nanometers into the clad region. To protect this field from interacting with foreign substances, the cladding layer is made several micrometers thick. The clad must also be transparent. Otherwise, it will absorb some of the evanescent tail, taking energy out of the propagating rays and significantly increasing attenuation.

In addition, an optional buffer layer (a sheath) may be placed around the cladding. This coating, which is often a colored organic layer, has no primary optical purpose; it is used to mechanically protect the glass core and clad from scratches and abrasion that would weaken the fiber.

4.3.1 Multimode Optical Fibers

In a multimode fiber, the rays within the TIR cone in Figure 4.12 comprise various propagating modes of the fiber (similar to the waveguide modes described in Section 3.5.3). Each mode propagates at a unique angle that is slightly different from the neighboring modes. In a typical multimode fiber, however, there are so many modes (thousands) that they appear to be almost continuous in angle.

How many of the possible modes are actually excited in a given fiber depends upon the input conditions. Light is usually coupled into the front face of the fiber from a source such as a laser, a light-emitting diode (LED), or a white-light source like a quartz-halogen incandescent bulb. The beam from a laser is very directional (narrow in angle), so only a few of the allowed fiber modes may be excited at the input. After propagating a distance in the fiber, however, these modes will gradually couple to other angles by scattering and bending of the fiber, and the excited mode structure will start to fill in. When the source is an LED or a bulb, the light emitted from the source is already broad in angle, and usually all the possible modes are excited at the input to the fiber.

The light that radiates from the exit end of the fiber (when the fiber mode content is full or almost full) comes out as a cone of light that can be related to the cone of modes inside the core. This situation is depicted in Figure 4.13. The cone of angles inside the core is broadened somewhat according to Snell's law of refraction as it exits the flat face of the fiber into the air. The half angle of the radiated cone in air is given by

$$\sin \phi_r = \sqrt{n_1^2 - n_2^2} \tag{4.15}$$

This radiation cone angle can also be put in terms of the numerical aperture (NA) of the fiber. The numerical aperture is defined as

$$NA = \sqrt{n_1^2 - n_2^2}$$

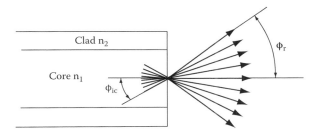

FIGURE 4.13
The exit conditions of rays radiating from the flat end of a multimode fiber. ϕ_r is the half angle of divergence, assuming all modes (i.e., all allowed propagation angles inside the core) are excited by the source.

so Equation 4.15 can also be written as

$$\sin \phi_r = NA \qquad (4.16)$$

For example, if $n_1 = 1.55$ and $n_2 = 1.51$, the numerical aperture of the fiber is $NA = 0.35$ and the half angle of divergence is 20.5° in air. The full divergence angle is therefore 41°.

BIOMEDICAL APPLICATIONS OF OPTICAL FIBERS

The largest use of optical fibers today is in telecommunications—most of your telephone calls and computer hookups are over fibers. Due to speed and distance requirements, almost all of these telecommunication fibers are single mode. In biology and medicine, however, most fibers are multimode since they are usually designed to carry light with high collection efficiency. Here fibers find three main applications:

- Endoscopes are image-transmission devices for peering inside the body. Bundles of thousands of small, closely packed multimode fibers are used to transmit the image. Each fiber transmits the intensity of one small segment (pixel) of the image. Some additional fibers in the bundle are used to carry illuminating light from the outside to the internal location. Since fibers can be bent, the endoscope is often flexible.
- Fibers are used to deliver laser energy to tissues for laser surgery in dermatology, in photodynamic cancer therapy, and other power delivery applications. Again, the fibers are flexible for convenient application. Included in this category are fibers for delivery of optical power for ablation of arterial plaque or for other tissue ablation. When the laser is a high-power CO_2 laser (10.6 μm infrared wavelength), the fibers must be fabricated from special infrared-transmitting materials because glass strongly absorbs at this wavelength.
- Fiber-based sensors for blood gases, blood proteins, and physical parameters such as pressure and temperature are being developed for *in vivo* measurements. The fibers allow direct placement of the sensors in tissues, veins, and arteries. One of the major advantages of fiber for biological sensing applications is that electromagnetic fields do not interfere with it. Thus, fiber optic probes are often used for temperature measurement in radio frequency (RF) heating applications.

4.3.2 Single-Mode Optical Fibers

The core diameter of a single-mode fiber is about 6 to 8 μm, and is much smaller than that of a multimode fiber. This means that only one mode will propagate in the fiber; all others are cut off, analogous to the cutoff condition already seen for microwave waveguides (Section 3.5.3.2). When there is only one mode, the ray picture used for multimode fibers is not adequate, and a full wave analysis must be used. The **E** field profile for a single-mode fiber is shown in Figure 4.14. There is a central peak in the **E** field magnitude, similar to that of the TE_{10} mode of a microwave waveguide (see Figure 3.41). However, in the case of a glass fiber, the **E** field does not go to zero at the dielectric interface between core and clad, as it must at the metal wall of a microwave waveguide. There is some penetration of the

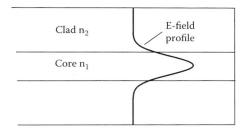

FIGURE 4.14
The **E** field pattern for the lowest-order mode (the only allowed mode) of a single-mode fiber. The core diameter is very small (typically about 6 to 8 μm). The mode shape can be approximated by a gaussian profile.

field into the clad region. This is the evanescent tail discussed earlier. Another difference is that the lowest-order mode of the fiber is cylindrically symmetric.

The radiation from the end of a single-mode fiber does not follow the same equation as for multimode fibers since there is no cone of rays in the single-mode fiber. Instead, the mode structure for the single-mode fiber in Figure 4.14 closely resembles a gaussian profile, which will be covered in Section 4.4, that stretches somewhat beyond the core (perhaps 20 to 50% depending upon the fiber parameters). Thus, the radiation angle is similar to that for a gaussian beam whose radius is approximately 1.2 to 1.5 times the core radius. The discussion of this radiation angle is deferred until Section 4.4.3; it will be shown there that for typical values, the divergence from a single-mode fiber is less than from multimode fibers.

4.4 Propagation of Laser Beams

As mentioned in the introduction, ray tracing is a very useful tool when diffraction is negligible. But when the beam size is small—as it is when generated from inside the active region of a laser—diffraction must be taken into account and ray tracing is not appropriate. This section discusses the kind of beams that radiate from lasers, and how they propagate after leaving the laser.

4.4.1 Linewidths of Laser Beams

One key characteristic that differentiates a laser beam from the beams that come from incandescent bulbs or LEDs is that its wavelength spectrum is much narrower. For optical sources, the wavelength spread is called *linewidth*. While a typical LED may have a linewidth that spans 30 to 40 nm, the spectrum of a laser diode (a semiconductor laser) may span only a fraction of a nanometer, depending upon the number of cavity modes present. Gas lasers (such as the HeNe laser) are even narrower in linewidth. Therefore, the laser beam has a more precisely defined wavelength than the other sources. It is said that the laser has a higher degree of temporal coherence.

Figure 4.15 shows the output spectrum of a typical visible laser diode. In the case shown, the laser has two modes, each at a unique frequency stemming from the resonance of the optical cavity that makes up the laser diode (analogous to the resonant wavelengths of a microwave cavity, discussed in Section 3.6). However, the number and position of the modes of a laser diode will often vary in time, especially if the temperature of the diode is not kept constant. This phenomenon is known as *mode hopping*.

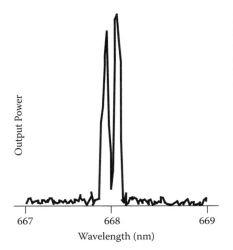

Output Power

667   668   669

Wavelength (nm)

FIGURE 4.15
Example of the output spectrum from a visible laser diode. Two modes are evident here, but some laser diodes have only one (single-mode lasers). The number, strength, and wavelength location of the longitudinal modes vary with temperature and drive current.

4.4.2 The Gaussian Spherical Profile

Another key characteristic of a laser beam is that it is generated by stimulated emission that builds up inside the optical cavity (a multiple-pass cavity known as a Fabry-Perot cavity). Because the beam bounces back and forth many times inside the cavity in the process of building up, it takes on a special profile. Only those beam profiles that reinforce themselves upon multiple bounces, that is, that keep the same shape after propagating back and forth several times, will survive in the laser's output. One profile that meets this requirement is the important *gaussian spherical beam*, sometimes simply called a gaussian beam.

The name *gaussian* refers to the shape of the beam's amplitude in the transverse direction. A gaussian shape means that the **E** field amplitude follows a quadratic exponential profile centered on the beam axis:

$$E = E_0 \, e^{-[r/w(z)]^2} \tag{4.17}$$

where r is the distance from the axis, E_0 is the magnitude of the **E** field on the axis, and w(z) is a parameter that specifies the radius of the beam at a distance z from a landmark location called the *waist*. At the waist, z = 0, w = w_0, and the beam width here is the smallest it gets (thus the name *waist*) until it is refocused by lenses. This shape is shown in Figure 4.16. Since a gaussian amplitude does not have sharp edges, the definition of width is somewhat arbitrary. By convention, the beam radius is defined as the point at which the field amplitude has fallen to 1/e (0.37) of its maximum value at the axis; this occurs at r = w(z). Since power density is proportional to the square of the **E** field, the power density will have fallen to $1/e^2$ (0.14) at this radius.

Note that the *diameter* of the beam is twice the radius, or d = 2w(z). Also note that the gaussian shape is an idealized approximation. Mathematically the **E** field amplitude never quite gets to zero even for very large distances away from the axis. In practice, however, the amplitude must go to zero (or small enough to be insignificant) at some finite distance. The gaussian approximation is sufficient because the energy carried in the extreme tail of the gaussian profile is small, so very little is lost if the profile is truncated at a reasonable distance from the axis.

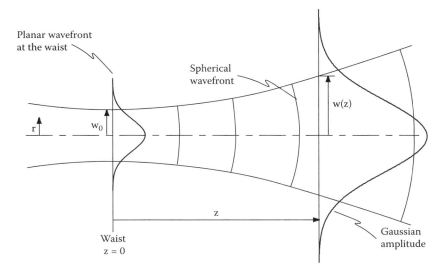

FIGURE 4.16
The propagation of a gaussian spherical beam away from its waist. The amplitude of the beam at any location z obeys a gaussian profile with radius w(z). Its wavefront is spherical with varying degrees of curvature. The propagation characteristics are symmetrical in z about the waist; that is, the converging beam to the left of the waist follows the same shape as the diverging beam to the right.

The word *spherical* is used to describe the curvature of the wavefront of the beam, which takes on a spherical shape that changes as the beam propagates. The wavefront curvature is simply another way of specifying the phase of the **E** field of the beam across any plane perpendicular to the direction of propagation. In Figure 4.16 it is seen that in general the wavefront is curved away from the waist, but at the waist the wavefront is planar (flat). In fact, the two distinguishing characteristics of the waist are: (1) the beam is smallest here, and (2) the wavefront is planar here. Where the waist actually occurs in space depends upon what components, such as curved mirrors or lenses, are changing the wavefront curvature of the gaussian beam; this is covered in more detail in the next two sections.

4.4.3 Propagation Characteristics of a Gaussian Beam

The parameter w(z) defines the radius of a gaussian beam at any location. As the beam propagates, w(z) will change. It is important to determine how fast w(z) changes, because this will determine how fast the beam spreads out (diverges) due to diffraction, or how much the beam will converge when focused. In this section, we look at diffraction of a beam propagating away from its waist; in the next section, we look at focusing by a lens.

To examine the propagation of a laser beam, it is first necessary to determine where the waist is located. For almost all lasers, the initial waist is located at or inside the laser cavity, either near the middle or at one of the end mirrors of the cavity. Each laser design will be slightly different. For example, a gas laser will usually have its waist at the output mirror of the laser. A semiconductor laser diode (whose oval output beam is roughly approximated by a gaussian beam with different waist sizes along each of two axes) will have its waist at the exit facet of the laser.

Once the waist is located, the beam radius can be found by noting that the beam expands monotonically as a function of the distance z away from the waist according to the following formula:

$$w(z) = w_0 \sqrt{1 + \left(\frac{z}{z_R}\right)^2} \qquad (4.18)$$

where z_R is the *Rayleigh range* defined by

$$z_R = \frac{\pi w_0^2}{\lambda} \qquad (4.19)$$

An example of this beam divergence is shown in Figure 4.17. Note from Equation 4.18 that the beam has expanded to just $\sqrt{2}$ times its waist radius when it reaches the Rayleigh range at $z = z_R$. The Rayleigh range is therefore one measure of the extent of the near field of the beam divergence.

Perhaps the most convenient way of expressing the spreading of the gaussian beam is to specify its far-field divergence angle. Note from Figure 4.17 that in the far field (i.e., more than a few Rayleigh ranges away from the waist), the beam edges—as defined by $w(z)$—are expanding linearly as a function of the distance z from the waist. This means that the beam looks like it is expanding linearly from a point located at the center of the waist. The divergence half angle ϕ_d, shown in Figure 4.17, can be obtained by letting $z \gg z_R$ in Equation 4.18 and using $\sin \phi_d \approx \tan \phi_d = w(z)/z$. This gives the divergence half angle as

$$\sin \phi_d = \lambda / \pi w_0 \qquad (4.20)$$

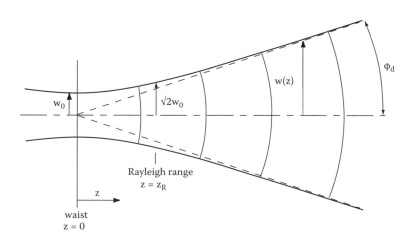

FIGURE 4.17
The far-field divergence half angle ϕ_d of a gaussian beam propagating away from its waist. The waist is a landmark feature whose radius w_0 determines how the beam will propagate. The location and radius of the waist are set by optical components forming the beam, such as laser cavities or intervening lenses.

Therefore, the smaller the waist size, the larger the divergence angle. This is consistent with the behavior seen earlier in the diffraction of a wave after passing through a small slit (Section 3.8.1).

The curvature of the wavefront also changes as the beam propagates away from the waist. At the waist, the wavefront is planar. As it propagates, it begins to curve slightly outward, and then becomes more curved at farther distances. In the far field, the wavefront curvature has a radius that is approximately equal to z, the distance from the waist. Thus, the beam approximates a segment of a spherical wave that appears to be emanating from a point centered at the waist. At very far distances, the spherical wavefront begins to look planar.

4.4.4 Focusing a Gaussian Beam with a Lens

When a gaussian beam is passed through a lens, a new waist is formed at the focal plane of the lens. This occurs because the curvature of the lens surface modifies the wavefront curvature of the beam by refraction, causing it to converge to a new waist location. (The radius of the beam is not changed immediately when passing through the lens, assuming that the lens diameter is somewhat larger than the beam size so that the beam is not clipped.) The diameter of the new waist is then a measure of the focused spot size.

Figure 4.18 shows the focusing configuration. The diameter of the focused spot is twice the new waist radius, or $d_0 = 2w_0$, and is given by

$$d_0 = 1.27 \, \lambda f / D \qquad (4.21)$$

where f is the focal length of the lens, λ is the wavelength, and D is the diameter of the beam as measured at the lens location. Thus, to make a focused spot smaller, a shorter focal length lens or a larger diameter input beam should be used. However, as mentioned above, the lens diameter must be greater than the beam diameter to avoid clipping, and it is difficult in practice to fabricate a lens whose diameter is much greater than its focal length; otherwise, aberration will be severe. In other words, f/1 is about the smallest f-number possible in a practical lens. Therefore, the smallest value the term f/D in Equation 4.21

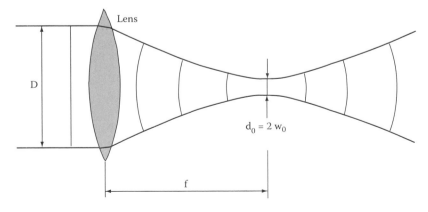

FIGURE 4.18
Focusing of a gaussian beam to a waist by a lens. As opposed to the simple ray picture, where the focused spot can be infinitely small, the actual spot diameter d_0 is finite. d_0 is a function of the beam diameter D, the focal length f, and the wavelength λ, as given by Equation 4.21.

can have in practice is approximately unity. In this case, d_0 will then be on the order of the wavelength of the light.

This illustrates a rather universal principle: about the smallest an electromagnetic wave can be focused is on the order of the wavelength of the wave. This sets the ultimate limit on the resolution of conventional optical microscopes, and is why short-wavelength light is preferred when a small focused size is needed to increase spot density in such applications as CD disks or optical storage. It is also why x-rays, with extremely short wavelengths, have such good spatial resolution when used to image tissues.

4.4.5 Applying the Gaussian Beam Equations

The gaussian beam equations of this section are a very good approximation to the beams from many lasers, such as gas or solid-state (e.g., NdYAG) lasers. For example, the beam exiting a typical HeNe laser has a waist radius of about 1 mm. Since $\lambda = 633$ nm, Equation 4.20 predicts a divergence half angle of only 0.011° (the full angle would be twice this value). This small amount of divergence is what makes a laser beam look like a pencil beam.

The gaussian equations do not apply quite as well to semiconductor laser diodes, since the typical edge-emission laser diode has an emitting junction that is a tiny, flat rectangle rather than a round mirror, but they still can be used approximately, with more accuracy in the short-axis dimension of the rectangle than in the long-axis direction. Because the emitting face is rectangular and because divergence is inversely proportional to the emission dimension (as seen in Equation 4.20), there will be more divergence in the laser diode's output beam in the plane perpendicular to the flat junction than in the plane parallel to the junction. As shown in Figure 4.19, this results in the output beam becoming oval, with the long axis of the oval rotated 90° with respect to the long axis of the junction.

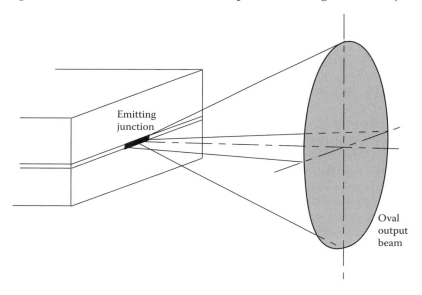

FIGURE 4.19
The output beam of a semiconductor laser diode. Because the emitting junction is rectangular, the divergence angles along the two transverse beam axes are unequal, with the divergence in the plane perpendicular to the junction's small dimension being larger than along the other axis. This leads to an oval output beam, unless corrected by special optical components.

As a numerical example, the short-axis dimension of the emitting junction of a typical laser diode may be only about 1.5 µm high. Letting $w_0 = 1$ µm (since the **E** field extends slightly beyond the confines of the junction) and $\lambda = 660$ nm in Equation 4.20 gives a half angle of divergence in the plane perpendicular to the junction of 12.1°, for a full angle of 24.2°. The long-axis dimension of the junction may be 10 µm wide. Assuming a gaussian profile along this dimension, Equation 4.20 predicts that the divergence half angle in the plane parallel to the junction would be 2.4°, for a full angle of 4.8°. Actually, because the **E** field profile in the long-axis direction is not really gaussian but is composed of higher-order transverse modes that are flatter in profile, the true divergence angle in this plane is somewhat larger than predicted by gaussian theory. The equations match closer along the small-axis direction than in the long-axis direction.

The divergence angles from an unfocused laser diode are rather large, as calculated above. To make the beam more directed, a short focal length lens is often placed close to the junction to collimate the beam. The new divergence angles are based upon the larger waist sizes at the lens, and divergence can therefore be much reduced.

Also, as mentioned in Section 4.3.2, the beam radiating from the face of a single-mode fiber can be approximated quite closely by a gaussian beam. In this case, the waist is located at the flat exit face of the fiber with a waist diameter approximately 1.2 to 1.5 times the core diameter. If the core diameter is 6 µm and $\lambda = 0.7$ µm, Equation 4.20 gives a divergence half angle of about 3.2°. Again, this can be reduced by a collimating lens.

4.5 Scattering from Particles

Small dielectric particles suspended in an aerosol or in a liquid will scatter electromagnetic waves in a manner that depends on the ratio of particle size to wavelength. For very small particles, such as those that will stay suspended for a long time in an aerosol or liquid solution, the sizes are very small—on the order of nanometers to micrometers. Therefore, typical sizes of suspended particles are often smaller than the wavelength of visible light. (The reason scattering is discussed in this chapter even though the wavelength is not much smaller than the particles is that most scattering of interest involves optical waves.) In other cases, the particle size can be comparable to a wavelength or larger.

The mechanism for the scattering of electromagnetic waves from particles is similar to what we encountered in Section 1.6 describing the interaction between **E** fields and dielectric materials. When an incident wave passes through a particle, the **E** field causes the electric dipoles in the particle's material (either induced or already existing) to align and alternate with the field, or, in conducting particles, causes free electrons to oscillate back and forth at the same frequency as the incident field. These oscillating charges act as small antennas, reradiating a wave that becomes the scattered wave.

The pattern of the scattered wave from this particle depends upon the relative phases of the wavelets emanating from the various portions of the particle. Thus, the pattern is sensitive to the size and shape of the particle. If the particle is very small compared to a wavelength, its radiation pattern falls into the classification of *Rayleigh* scattering. If its size is on the same order as a wavelength, the pattern is much more complex and the scattering is known as *Mie* scattering. The characteristics of these two scattering regimes are discussed in more detail in the following sections.

Of course, there are many particles in the scattering cloud, so the individual waves from each particle combine to form the entire scattered wave. In most cases of interest, the individual particles are randomly located in the cloud, so the waves from these particles have random phases uniformly distributed over 0 to 360° when they reach the observation point. When their **E** fields add, the resultant total **E** field has a magnitude whose average is statistically equal to the square root of the sum of all the individual **E** field magnitudes. The power density in the scattered wave, proportional to the square of the resultant **E** field amplitude, is therefore equal to the sum of the power density scattered from each particle. This is an example of the *incoherent* addition of powers from individual particles. If there are N particles in the cloud, and each particle scatters P_p power, the total average power in the scattered wave is $P_{total} = N \times P_p$.

4.5.1 Rayleigh Scattering

When the particle is small compared to a wavelength, the scattering contributions coming from each segment of the particle are approximately in phase. Thus, the particle acts as a single small dipole antenna. We have seen earlier (Section 3.7) that the radiation pattern from a small dipole is uniform in angle in the plane perpendicular to the **E** field, radiating equally in all directions in that plane, but falls off in the direction parallel to the **E** field vector in the plane containing **E** (as shown in Figure 3.51 earlier). The Rayleigh scattering pattern follows that same donut-shaped behavior, as indicated in Figure 4.20: it scatters uniformly in the forward, sideways, and backward direction, decreasing only in the direction aligned with the incident electric field.

The magnitude of the scattering from a Rayleigh particle depends strongly—in fact, to the fourth power—on the size of the particle compared to the wavelength. If the effective

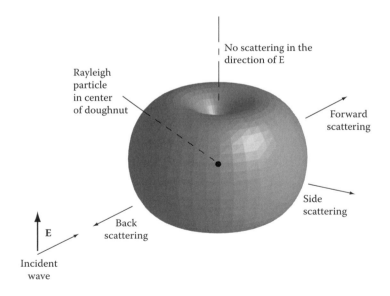

FIGURE 4.20
A polar plot of the scattering efficiency of a small particle. The particle's size is much smaller than the wavelength of the incident light, so its scattering behavior follows Rayleigh scattering equations. Note that scattering is uniform in the plane perpendicular to the incident **E** field vector, but falls to zero in the direction parallel to the **E** field. The polar plot therefore has a doughnut shape.

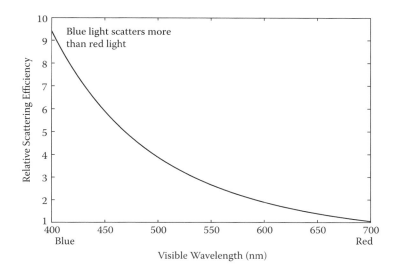

FIGURE 4.21
The relative scattering efficiency of a Rayleigh particle of a given size for different visible wavelengths. Due to the fourth-power wavelength dependence, blue light is scattered much more than red light.

particle size is s, the power scattering efficiency of the particle is proportional to $(s/\lambda)^4$. Over the visible wavelengths, there is a dramatic change in the scattering power for a particle of a given size when going from the short-wavelength end of the spectrum to the long-wavelength end. Figure 4.21 shows that the relative scattering efficiency is more than nine times larger for blue light than for red light. This helps explain the blue color of the sky. We see scattered sunlight when we look up through the atmosphere, and this scattering (from the molecules in the air) is much more effective for the blue portion of sunlight than for red. Even though the scattering from each air molecule is very small, the huge number of molecules makes the total scattered power visible. Random fluctuations in the air density also cause scattering with the same behavior.

The $(s/\lambda)^4$ scattering dependence that is characteristic of Rayleigh scattering also applies for a fixed wavelength as the particle size varies. Figure 4.22 shows the very large variation in relative scattering efficiency for a fixed wavelength as the particle size varies over just a tenfold range. Obviously, larger particles are much more efficient light scatterers than small particles in the Rayleigh regime.

4.5.2 Mie Scattering

For larger particles, ones whose sizes are on the order of the wavelength of light, the scattering pattern and scattering efficiency become much more complicated than the rather uniform pattern of the Rayleigh particle. Since the contributions to the scattered wave emanating from different parts of the particle are spread over distances that are a significant fraction of a wavelength (or even a few wavelengths) apart, the large phase variations of these contributions cause major constructive and destructive interference that is a sensitive function of the angle, shape, orientation, refractive index, and conductivity of the particle. Thus, the pattern of Mie scattering can be very irregular, with numerous peaks and valleys at various angles. For example, Figure 4.23 shows a polar plot of the Mie scattering in the plane perpendicular to the incident **E** field for a spherical particle that is

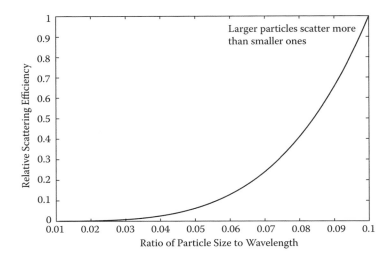

FIGURE 4.22
The relative scattering efficiency of Rayleigh-sized particles of varying sizes as a ratio to wavelength. The fourth-power dependence that is characteristic of Rayleigh scattering makes larger particles scatter considerably more efficiently than smaller particles.

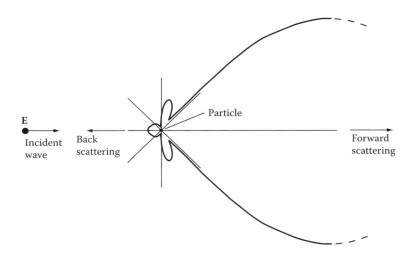

FIGURE 4.23
A polar plot of scattering from a Mie-sized spherical particle. This plot shows scattering efficiency as a function of angle in the plane perpendicular to the incident **E** field. In this example, the particle's diameter is 1.27 times the wavelength and the particle's refractive index is 1.25. Note the large forward scattering component characteristic of Mie scattering, as opposed to the uniform scattering expected in this plane for much smaller Rayleigh particles.

1.27 wavelengths in diameter with a refractive index of 1.25. Even this simple spherically shaped particle displays an irregular scattering pattern.

Although the Mie patterns are relatively complex, there is one characteristic that is commonly seen in the patterns from spherical and similarly shaped particles. The forward-scattered power, which is the scattering into a cone in the same direction as the incident wave, is greater than the scattered power in other directions (this is known as the *Mie effect*). The forward scattering component becomes even more pronounced as the particle diameter increases.

When the particle size gets very large (much larger than the wavelength of the light), geometrical optics becomes the applicable analysis tool and ray tracing can be used, as discussed in the first part of this chapter.

SCATTERING FROM PARTICLES

You can easily demonstrate the effects of light scattering from particles in your lab or at home by passing a light beam (from a laser pointer or a flashlight, for example) through a solution of scattering particles such as a glass of milk that has been diluted amply with water so the beam passes through several centimeters without suffering too much attenuation. In a darkened room, observe the brightness of the scattering as a function of angle (but be careful not to look directly into the transmitted laser beam!) to see the increase in scattering in the forward direction compared to the side and back directions.

If you are using the white beam from a flashlight, notice also that the color of the forward-scattered light takes on a reddish or yellowish tint. This is due to the higher amount of scattering of the blue light out of the way from the red light, for the reason mentioned above. (This is also the source of beautiful red sunsets, accentuated when there is an excess of haze or pollution in the air.) If you are using a polarized laser, notice that the scattering off to the side in the direction aligned with the incident **E** field is lower than that in the perpendicular plane, as given in Figure 4.20 for Rayleigh particles. This characteristic is seen in smaller-sized Mie particles as well as Rayleigh particles.

The increase in both Rayleigh and Mie scattering power when the particle size gets larger sometimes has unfortunate consequences in human vision as a person gets older. With aging (and exposure to ultraviolet rays), some people will develop cataracts in the lenses of their eyes. These cataracts are caused by aggregation of proteins in the lens, producing density variations with corresponding refractive index variations. These result in small scattering centers. As the cataracts grow, they scatter more light out of the normal path to the retina, resulting in dimmer and blurred images. When severe enough, the cataracts have to be surgically removed to restore reasonable vision. However, even before they cause noticeably blurred images, developing cataracts can cause scattering. This is especially evident as "halos" around a bright light source (forward scattering) that can obscure the images of nearby objects, and many older people complain of this halo effect when driving at night in the presence of oncoming headlights.

4.6 Photon Interactions with Tissues

It is obvious by visual inspection that the tissues of the body have much different optical properties than typical dielectric materials used in optics, such as glass. For example, consider what happens when a bright beam of light (from a flashlight or laser) is shined through your closed fingers and observed from the opposite side in a darkened room. Even though the beam may be fairly directed (collimated) when entering the tissue, by the time it exits it has spread into a large glow. It also has lost some of its intensity by absorption. This illustrates two main characteristics of the optical properties of tissues (as represented here by soft tissue): First, there is a large amount of scattering by the microstructure of the tissue. Second, there is a fair amount of absorption, which in many tissues is wavelength dependent, and which therefore imparts a color to the tissue. We cover these aspects in more detail in the next two sections.

4.6.1 Light Scattering in Tissues and Photon Migration

A simplified picture of how light propagates through tissue is offered in Figure 4.24. This shows the paths that a light beam might take as it passes through the tissue. The most obvious feature is the high degree of multiple scattering that occurs as the light encounters numerous layers of inhomogeneous tissue components and multiple sites of scattering. The multiple scattering is indicated by the zigzag path of the light rays. Upon each scattering event, the direction of the light is scrambled.

The use of rays here to describe the light propagation is not strictly valid, but is used as a visualization tool. An alternative description is to consider the paths that photons (the quantum-mechanical particles that make up the energy of a light beam) would take as they transit through the tissue. The paths may look similar to the zigzag paths of Figure 4.24. This viewpoint, known as a study of *photon migration* in tissue, is a rich research topic. Statistical estimates of the propagation of photons through various types of tissue can be made from this viewpoint. Many of these studies use the so-called Monte Carlo computer technique, where a vast number of individual photons are tracked one at a time through the tissue, each one undergoing a large number of random scattering events; then the paths are added up to arrive at an estimate of the total effect on the light beam.

An estimate of the degree of the multiple scattering can be obtained from the value of the scattering coefficient μ_s' of the tissue. For typical soft tissue, μ_s' is approximately in the range 0.5 to 4 mm^{-1}. In rough terms, this means that a photon will travel only a distance of $1/\mu_s' = 0.25$ to 2 mm before encountering the next scattering event. Even in a small volume of tissue, there are many scattering events scrambling the path of the light.

The high degree of multiple scattering in tissue makes ordinary optical microscopy difficult at any depth below the surface of the skin. It is like trying to see objects through a very dense fog. There is, however, a new development in tissue microscopy, called optical coherence tomography (OCT), which uses the partial temporal coherence properties of nonlaser sources to image objects at moderate distances (a few millimeters) beneath the surface of the skin.

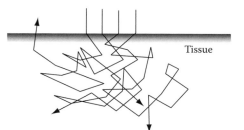
Tissue

FIGURE 4.24
A simple picture showing the paths that photons might take in traveling through soft tissue. Note the high degree of multiple scattering, making imaging with light difficult at any depth beneath the skin's surface.

4.6.2 Tissue Absorption and Spectroscopy

In addition to being highly scattering, typical tissues (other than the humour of the eye) have a moderate amount of absorption. The magnitude of tissue absorption may be estimated from the absorption coefficient μ_a, which for soft tissues is approximately in the range 0.01 to 1 mm^{-1} (there is a large amount of variation between tissue types and with wavelength). This means that a light beam will travel approximately $1/\mu_a = 1$ to 100 mm before losing an appreciable amount of its energy by absorption. Note that the absorption coefficient is smaller than the scattering coefficient in the previous subsection, which means that a photon on average will encounter many scattering events before being absorbed or exiting out from the tissue.

One interesting use of optical absorption in tissues is for diagnostic purposes. Here the wavelength dependence of the absorption is employed as a means of measuring the state of the tissues, either for detecting a disease condition or for monitoring the local environment of the tissue. This is using *spectroscopy* for diagnostics or sensing. For example, it is well known that when oxygenated, blood has a red color; when the blood oxygen is depleted, it takes on a blue color. In reflected light, arteries look reddish and veins look bluish. The reason for this is that oxygen is carried predominantly by hemoglobin molecules found in red blood cells. When hemoglobin is oxygenated (oxyhemoglobin, HbO_2), it absorbs red light to a lesser degree than when it has lost oxygen (reduced to hemoglobin, Hb).

This characteristic is shown in Figure 4.25, which is the absorption curve for both states of hemoglobin as a function of wavelength. Note that at a red color (say $\lambda = 660$ nm), the absorption is less for oxyhemoglobin than for deoxygenated hemoglobin. Thus, red light from an LED passing through perfused tissue will undergo a variable amount of absorption depending upon the degree of oxygenation of the blood. This allows the measurement of the oxygen saturation of the blood with relatively simple devices known as pulse oximeters, which clamp on the finger, toe, or earlobe. In-dwelling fiber optic oximeters also have been developed for insertion directly into an artery. Usually two or more wavelengths of light are used to account for the variability in tissue other than that caused by the concentration of oxygen. For example, Figure 4.25 shows that at a wavelength of about 805 nm, absorption is relatively unaffected by the degree of oxygenation; this is an *isosbestic* point. The ratio of absorption at these two wavelengths, 660 and 805 nm, will remain sensitive to oxygenation while being less sensitive to other factors.

There are other spectroscopic uses of light for monitoring the state of tissues in the body. Several of these are currently being investigated for their clinical usefulness. Fluorescence induced in various tissues (autofluorescence) by an external light source such as a laser can yield information about the nature of the tissue, such as discriminating calcified plaque from

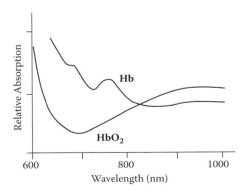

FIGURE 4.25
The absorption characteristics of oxyhemoglobin (HbO_2) and hemoglobin (Hb) as a function of wavelength. Oxyhemoglobin absorbs less red light than hemoglobin, so oxygenated blood looks redder than deoxygenated blood. This effect is used in pulse oximeters, which use red light to noninvasively measure the oxygen saturation of a patient. A second (or more) wavelength at the isosbestic point—where absorption is not sensitive to oxygen percentage, near 805 nm—is often employed to compensate for other optical variables.

soft plaque inside arteries. Optical fibers carry the laser light to the tissue and collect the emitted fluorescence. Raman scattering, which is very specific to the tissue type but which is also very weak in signal strength, is also being investigated for diagnostic applications.

INFRARED THERMOMETRY

Infrared radiation is given off by warm objects. Infrared sensors are robust, accurate (to better than a tenth of a degree), and relatively inexpensive, and are used for numerous applications to measure temperature without actually touching the object to be measured. Since biological processes produce heat (and therefore emit infrared radiation), many aspects of the health of the body can be monitored using infrared thermometry. Figure 4.26 is an infrared photograph (thermograpic image) of a hand taken with an infrared camera. Perhaps the most widespread application of this technology is the noncontact ear thermometer. The ear drum provides an excellent window to the core body temperature. Infrared sensors can read its temperature without making contact with its sensitive structure.

Infrared scanning has also been used to measure the temperature in the body during hyperthermia treatment. But due to the limited depth of penetration of infrared wavelengths in tissue, the temperature maps are of the surface only, and surface temperatures are often a poor indicator of temperatures deeper in the body. This is also true of using infrared thermometry for breast cancer detection.

During heating device development and testing, however, it is possible to create a semisolid or rigid phantom that opens up like a clamshell to allow the infrared camera to capture the two-dimensional heat distribution pattern. This must be done quickly, of course, before the heat is conducted away, thus distorting the pattern.

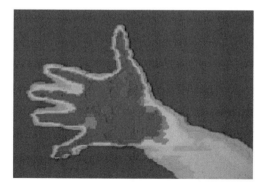

FIGURE 4.26
Please see color insert following page 146. Infrared image of a hand taken with a thermographic camera. (From M. Iskander. With permission.)

4.7 X-Rays

The previous sections have described ways in which electromagnetic waves act like rays that can reflect, transmit, and refract. They can be guided for use in lenses, mirrors, and optical waveguides. But they still have a fundamental similarity with the waves in the previous chapters: they act like waves (not particles). Going even higher in frequency, to the x-ray frequencies, the electromagnetic waves begin to behave very differently. They begin to act more like particles than waves, and their effect on the body is very different as well.

Modern physics tells us that all waves can be modeled as either waves or particles, depending on their energy relative to the energy of their nearby environment. X-rays are considered to be particles (photons) rather than waves because when they hit a molecule, they may break its molecular bonds or ionize the molecule. This is of particular concern with the DNA molecule, because it can now cause the molecule to mutate into a cancerous form.

The reason the higher-frequency waves can ionize atoms, whereas the lower frequencies do not, is that the total available energy of a photon, given by Planck's law, is inversely proportional to wavelength. The energy of a photon (in units of electron-volts, or eV) is 1.24×10^{-6} eV/λ, where λ is in units of meters. Approximately 10 eV is required to ionize a molecule. This corresponds to a wavelength of about 0.124 µm, which falls in the deep UV region. Particles with frequencies in the visible light spectrum and below are unable to break even the weakest molecular bonds. Thus, deep UV light and x-rays are treated as particles of ionizing radiation. Visible light, infrared, and frequencies used for communication and the medical applications described in this book are treated as waves and are nonionizing radiation.

4.8 Measurement of High-Frequency Electric and Magnetic Fields (Light)

At high frequencies, the electric and magnetic fields are strongly coupled. As with low- and mid-frequency fields, optical measurements typically take an analog measurement and convert it to a digital value, although it is common for preprocessing to be done with filters, mirrors, lenses, and so forth prior to digitizing the final signal. Light is typically measured according to its intensity (power density) using detectors such as the silicon pin photodiode for visible light; other semiconductor materials are used for other wavelengths. Phase is very difficult to measure, although diffraction gratings and cavities can be used to create interference patterns that can be used to measure phase and the effects of phase.

Laser beams and optical fibers are relatively easy to control spatially. Fiber optics, in particular, tend to be small, lightweight, biocompatible, and easy to use. They have found their way into a wide array of sensors based on the amount of power or the spectrum returned in the fiber optic cable. When used for imaging (such as for endoscopy), the image is originally analog but is converted to digital by devices such as the charge-coupled detector (CCD).

Grating and prism spectrometers are the equivalent of spectrum analyzers for light and can measure the pattern of the optical spectrum.

BIOMETRICS

A biometric system uses a person's unique biological attributes (fingerprint, DNA, vein patterns, and iris scans are the most prevalent today) to identify that specific individual. These systems can be used in two major ways. Authentication is used to determine that you are who you say you are, such as fingerprint scans used to log in to a computer. This is a type of a biological password. The other major way a biometric system is used is to identify an individual from a much larger group, as in the case of fingerprint matching to identify criminal suspects. Standard ink-based fingerprinting has been used for over a century. Originally this system required a human to match the fingerprints, but now computer scanners are used for the task.

New optical technology is expanding the biometric opportunities available, and hence the applications available. Iris scans are one such example. Each person's iris is as unique as his or her fingerprint, and an optical iris scan (a picture) can be used to uniquely identify an individual. This method is being considered for immigration and border control because it is noninvasive, quick, and relatively inexpensive. Another optical method for identifying people is through the shape of their hand and fingers. Most often used for simple access control (unlocking doors), this simple photo-based method is often considered less invasive to privacy than fingerprinting or iris scanning. This method has its limitations, however, as a person's hands change with time due to weight change, arthritis, or injury. Voice recognition is another identification method that has a similar limitation. Voice is typically used less than other biometrics, because of the possibility that a recording of that individual could be used to gain access. For the most secure applications, there is always concern that the biometrics could be mimicked using physical models of the person's hands, or even his or her fingerprint. Also, to avoid identification, these could be changed (such as by filing down the fingerprint ridges, for instance).

Infrared and near-infrared scans can be used to identify biological characteristics below the surface of the skin, thus making them nearly impossible to change or copy. Vein geometry is one biometric characteristic that is unique to each individual. Even identical twins have different vein patterns, and a person's veins vary from the left to right sides. The shape of a person's veins changes very little with age (although injury may have an impact on them). Many veins are not visible through the skin, making them extremely difficult to counterfeit or tamper with. A vein scan is quick and noninvasive. A picture is taken with near-infrared light (much like a photo, but with a different wavelength). The hemoglobin in the blood absorbs the infrared, as shown in Figure 4.25, but the rest of the body reflects it. Thus, the vein pattern appears black and can be compared to previously stored patterns.

4.9 Summary

High-frequency electromagnetic fields act for the most part like rays. Most medical applications in this band are in the optical range. There are many infrared applications, for instance. Infrared applications are often low cost and noncontact, making them particularly good as biological sensors. Applications such as spectroscopy are capable of evaluating biological tissues to determine chemical composition, for instance.

Applications in the high-frequency range are tending toward smaller and smaller devices, mainly sensors, many of which may become candidates for implantation in the body due to their small size. These are discussed in more detail in Chapter 6.

5

Bioelectromagnetic Dosimetry

5.1 Introduction

Dosimetry is used to determine the strength, direction, and polarization of electromagnetic fields. It predicts the *dose* of the electromagnetic field present at any point inside or outside of the body. For instance, it is used to predict the strength of fields in the head from cell phones to determine if a particular design meets regulatory guidelines, to determine the signal-to-noise ratio (and hence image quality) for coils for magnetic resonance imaging (MRI), to determine the field strength in experimental setups such as petri dish exposure systems, and many other applications.

The simplest form of dosimetry consists of two parts. The first is the determination of the *incident fields*, which are produced by some kind of source. These incident fields are either measured (with no object present) or calculated from a knowledge of the source. The second part is the determination of the **E** and **B** fields inside an object exposed to the incident fields. The fields inside an object are called the *internal fields*. The internal fields are also either measured or calculated. This type of dosimetry can be used when the object and source are far enough away from each other that the presence of the object does not change the configuration of the source. This is generally true for exposure systems, transverse electromagnetic (TEM) cells, and so forth. Low-frequency systems can also utilize this type of dosimetry.

For systems where the source is close to the object (such as cell phones), the presence of the object changes the incident fields from the source. This *coupling* must then be modeled within the dosimetry calculations. Methods that do this are sometimes referred to as *full-wave* simulations. Most mid-frequency systems require this type of dosimetry.

In early dosimetry models, spherical models were used to represent the biological model. Spheres obviously do not represent the shape of people or animals very well, but much useful information that served as a basis for calculations in more realistic models was obtained from spherical models. Subsequently, spheroidal (egg-shaped) and ellipsoidal models were used to represent animal shapes more realistically. Spherical, spheroidal, and ellipsoidal models were particularly useful because analytical, or closed-form, solutions could be obtained for them. That is, Maxwell's equations could be solved and mathematical expressions for the internal **E** and **H** fields could be obtained. Such solutions provided valuable understanding about the characteristics of internal fields.

Later, other models with more realistic shapes were used in both experimental measurements and calculations. A very common approach is to use *block models* of the body of interest. A block model consists of cuboidal mathematical cells arranged to approximate the shape of human and other animal bodies. Analytical solutions usually cannot be obtained for these models. Instead, *numerical methods* are used to calculate the internal fields. These numerical methods consist of solving Maxwell's equations using some kind of computer technique that gives the **E** and **H** fields in each mathematical cell of the model. In some of these numerical techniques, cubical mathematical cells are used; in others, pyramidal cells are used.

Some commonly used numerical methods are the moment method (MoM), finite-element method (FEM), impedance method, finite-difference time-domain (FDTD) method, and finite-difference frequency-domain (FDFD) method. In general, models with more realistic shapes require a larger number of smaller mathematical cells. With a larger number of smaller mathematical cells the fields can be calculated with finer resolution, but this also requires more computer memory and computational time. In recent years, computers have become more and more powerful, allowing more and more sophisticated dosimetry calculations. Calculating fields with more resolution is not enough, however. Having huge data files consisting of **E** and **H** fields in each of many mathematical cells does not necessarily provide insight into the characteristic behaviors of these fields. An understanding of this characteristic behavior is important in interpreting and evaluating the interactions of EM fields with biological systems.

Sometimes internal fields are measured in experimental animals, and sometimes they are measured in models consisting of material that has permittivity and conductivity similar to that of animal tissue. These models are called *phantoms*. Because measurements of internal fields in humans cannot be made, measurements in phantoms of humans are made to determine what the internal fields would be in the human body. An example of a phantom used to measure the power deposition in the head from a cellular telephone is shown in Figure 5.1. Appendix A describes some of the phantom materials available. When internal fields are calculated, various mathematical models are used to represent humans and other animals, as explained below. In bioelectromagnetics, information about the internal fields is usually desired so that effects of the internal fields on the biological system can be determined.

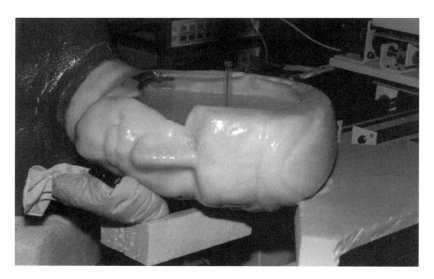

FIGURE 5.1
Please see color insert following page 146. Measurement of fields in a human head-hand-torso phantom from a cellular telephone. The head and torso are made from epoxy doped with salt to have the electrical properties of bone. The head is filled with semisolid phantom material that has the properties of brain. The hand is a rubber glove filled with material representing 2/3 muscle. The three components (x, y, z) of the electric field are being measured with an electric field probe similar to that described in Section 1.2. The specific absorption rate (SAR) described in Section 1.14 is calculated from these fields.

Dosimetry can be divided into two categories: *macroscopic* and *microscopic* dosimetry. In macroscopic dosimetry, the EM fields are determined as an average over some small volume of space, such as in mathematical cells that are centimeters or millimeters in size. For example, if the mathematical cell size is 1 mm on a side, then the **E** field in a given mathematical cell is assumed to have the same value everywhere within the 1 mm^3 volume of that cell. In other words, the **E** field is averaged over the volume of the cell. The **B** field is also averaged over the cell. These are called macroscopic EM fields. In contrast, in microscopic dosimetry, the EM fields are determined at a microscopic level, such as the cellular level in biological systems. Or, equivalently, the mathematical cells over which the EM fields are determined are microscopic in size.

Historically, much more has been done in macroscopic dosimetry than in microscopic dosimetry. Only recently has much work been done in microscopic dosimetry. One technique is first to determine the macroscopic fields, and then from the macroscopic fields, "zoom in" to find the fields on a microscopic level. Microscopic dosimetry is needed to learn more about how EM fields interact with biological systems at the cellular level, but both calculations and measurements are much more difficult for microscopic dosimetry than for macroscopic dosimetry. One obvious difficulty in microscopic dosimetry is managing the huge amount of data involved in systems that consist of millions of biological cells.

The purpose of this chapter is to discuss the principles and ideas involved in dosimetry. This is primarily used in two cases: when the wavelength is large compared to the object (see Chapter 2), and when the wavelength is about the same size as the object (see Chapter 3). When the wavelength is very small compared to the size of the object, the internal fields are confined to a very thin region near the surface of the object. In that case, dosimetry is rarely needed. Dosimetry is a complicated subject, and the literature describing the development and implementation of both theoretical and experimental dosimetric techniques and the resulting dosimetric data is extensive. The purpose of this chapter is not to give a comprehensive review or discussion of dosimetric techniques or to give a comprehensive set of data for the internal fields for various models of humans and other animals. Instead, we hope to help you understand the elementary ideas involved in dosimetry, and to give you a brief introduction to some of the techniques that are used in dosimetry. The first part of this chapter describes models available for dosimetric calculations. This is followed by a discussion of how the power deposition (specific absorption rate [SAR]) depends on the many parameters of the body and incident electromagnetic fields using an analytical solution and a spheroidal model. This analytical approach is good for providing an intuitive understanding, but not for providing precise results, particularly for localized SAR values. A later section describes how the finite-difference time-domain (FDTD) method can be used to determine the localized SAR values. FDTD is the most prevalent method for bioelectromagnetic dosimetry today. Examples of how the FDTD method is used to calculate the power distribution from cellular telephones and 60-Hz power lines show how this method is applied. The FDTD section is followed by a description of how the impedance method is used to calculate the fields around medical implants in the human body.

5.2 Polarization

The polarization of the electric field significantly impacts how it is absorbed and reflected, and how it is transmitted and received. As described in Section 1.2, the electric fields are often transmitted and received with wire antennas that are parallel to the electric field vector. This vector defines the polarization of the electric field. The magnetic field vector is then perpendicular to the electric field and to the direction of propagation of the wave, as described by the right-hand rule in Section 1.3.

The calculated **E** fields inside prolate spheroidal models in Section 2.3 illustrate the general characteristic that internal fields vary with the orientation of the incident fields with respect to the object. In addition to the polarization of the electric field relative to space (such as Ex, Ey, Ez polarizations), the orientation of the incident **E** field with respect to an object is also defined as *polarization*. For objects of rotation (those with circular symmetry about the long axis), three polarizations are defined, E, H, and K, as illustrated in terms of a prolate spheroidal model in Figure 5.2. E polarization is when the incident **E** field lies along the long axis of the object, H polarization is when the incident **H** field lies along the long axis, and K polarization is when the propagation vector **k** lies along the long axis of the object.

For objects that are not objects of rotation, such as the human body, six polarizations are defined, as illustrated in Figure 5.3 in terms of an ellipsoid. An ellipsoid has three semi-axes; the lengths are labeled a, b, and c in Figure 5.3, with $a > b > c$, and a lies along the x axis, b along the y axis, and c along the z axis. The polarizations are defined with respect

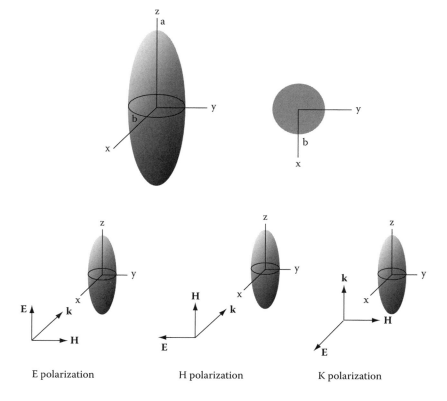

E polarization   H polarization   K polarization

FIGURE 5.2
Definitions of polarizations for a body of rotation about the z axis. For convenience, the polarizations are illustrated with respect to a prolate spheroid, but they apply to any body of rotation. (Adapted from Figure 3.37, Durney, C. H., et al., *Radiofrequency radiation dosimetry handbook*, 4th ed., Report USAFSAM-TR-85-73, USAF School of Aerospace Medicine, Aerospace Medical Division (AFSC), Brooks Air Force Base, TX, 1986.)

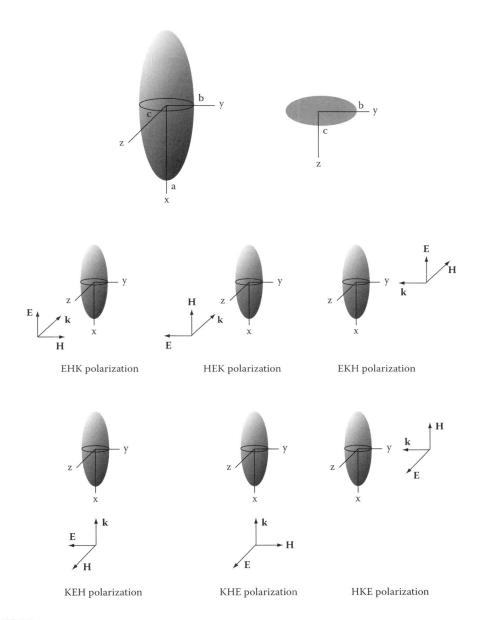

FIGURE 5.3
Definitions of polarizations for a body that is not a body of rotation. For convenience, the polarizations are illustrated with respect to an ellipsoid, but they apply to other objects as well. (Adapted from Figure 3.38, Durney, C. H., et al., *Radiofrequency radiation dosimetry handbook*, 4th ed., Report USAFSAM-TR-85-73, USAF School of Aerospace Medicine, Aerospace Medical Division (AFSC), Brooks Air Force Base, TX, 1986.)

to which of the vectors **E**, **H**, and **k** lie along which of the three semiaxes. For example, in EHK polarization, **E** lies along *a*, **H** along *b*, and **k** along *c*.

Another way that polarization can be defined is *horizontally* or *vertically*. This refers to the direction of the **E**-field vector relative to the orientation defined in the application. As before, the magnetic field vector follows the electric field vector and direction of propagation according to the right-hand rule.

Yet another aspect of polarization is how the electric field vector changes as it propagates with time. To determine this type of polarization, imagine the electric field vector propagating away from you. Watch the tip of the electric field vector as it propagates with time. If the vector traces out a line, then the wave is *linearly polarized*. If it traces out a circle, it is *circularly polarized* (or CP). And if it traces out an ellipse, it is *elliptically polarized*. Circular and elliptical polarizations are either left- or right-handed, depending on which direction the wave circulates as it propagates. With the thumb in the direction of the propagation, the fingers of the left or right hand, respectively, will follow the direction of rotation to determine the handedness of the wave. The time-dependent polarization of the wave is most important for communication applications. As waves reflect off multiple random scatterers, the wave generally becomes circularly polarized, so many communication applications utilize circular polarization. Virtually all dosimetry models to date have utilized only linear polarization. This is because they are generally used to determine the largest power deposition in the body, which can be found from aligning the long axis of the body (head to toe) with the electric field vector.

5.3 Electrical Properties of the Human Body

Both theoretical and experimental dosimetry require a knowledge of the electrical properties of the object for which dosimetry is being assessed. In particular, when the object is the human body, its electrical properties must be known to calculate internal fields or to construct phantoms on which measurements can be made. The electrical properties of the human body are usually specified either in terms of relative permittivity and effective conductivity, or in terms of complex relative permittivity (see Sections 1.6 and 1.14). In general, the relative permittivity and effective conductivity of body tissues are a strong function of frequency. Each of the different tissues in the body has a different variation with frequency. All of this must be taken into account in dosimetry. Appendix A gives more details on the electrical properties of many of the tissues in the body as a function of frequency.

5.4 Human Models

Model development is one of the significant challenges of numerical bioelectromagnetics for dosimetric calculations. Models have progressed from the prolate spheroidal models of the human used during the 1970s to roughly 1 cm models based on anatomical cross sections used during the 1980s to a new class of millimeter-resolution MRI-based models of the body, which are the hallmarks of research in the 1990s and beyond. These models can now be posed in virtually any position using computer graphics techniques. For instance, Figure 5.4 shows the "visible man" derived from MRI scans by the National Institutes of Health. This model is probably the most widely used of the voxel-based models available at this time. The scans were originally made with the arms lying at the sides; however, many applications (such as analyzing a man driving a car) require the arms to be in a different position. Figure 5.4 shows the arms in an upright position to hold on to a steering

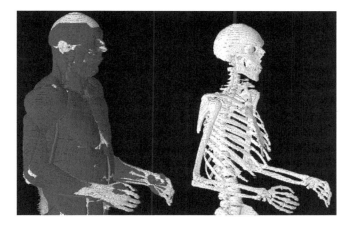

FIGURE 5.4
Please see color insert following page 146. The visible man from the National Institutes of Health was originally scanned with the arms at his sides. Here they have been computationally repositioned, with care being taken to ensure that the bones and other structures remain intact. (From Remcom, Inc. Reprinted with permission.)

wheel. The skeletal system is also shown. It is important when using a computer to reposition a model that the physical pathways for currents and fields be preserved. Thus, care must be taken to ensure that the bones and other structures remain intact during the repositioning process.

MRI scans provide an ideal initial database for voxel-based models of this type, but the scans alone do not define the types of tissue that are in each location. Instead, MRI scans provide a voxel map of MRI densities, which unfortunately do not have a one-to-one correspondence to tissue type. These images are interpreted as gray-scale images by which the several tissues can be seen. Image segmentation is necessary to convert these density mappings into mappings of tissue type. This is generally done semimanually, although automatic methods have been used as well.

Several MRI-based models of the human body, the head alone, and a wide variety of animal models exist, including the visible man shown in Figure 5.4. With the exception of some basic automatic tissue classification techniques based on MRI densities (dry tissue can be separated from wet tissue, for instance), these models have required significant effort to obtain, and there are many unique challenges in developing models suitable for use in bioelectromagnetic modeling.

First, there are issues that must be addressed in obtaining the MRI scans. It is important to use MR settings to optimize the contrast between the soft tissues, and to use saturation pulses to reduce pulsatile blood flow artifacts and time gating to reduce blurring from breathing and the beating heart. Depending on the amount of time gating and optimization, scanning the complete body with a vertical resolution of 3 mm takes 6 to 24 hours. The person being scanned will need to be repositioned during this time, as that is too long to expect a living person to hold still, and this presents some difficulties in rematching the images from successive positions. It is useful to position the person in the stature that is desired for modeling, such as ensuring that the feet are in a standing position, as opposed to relaxed, and that the head is in alignment with the spine, as opposed to on a pillow. Arms have caused significant difficulty in several modeling efforts, as in a relaxed position, they tend to fall out of the range of MRI scanning. Most of these problems are eliminated if a cadaver is used as the subject to be scanned, such as in the Visible Man Project, although the difficulty of positioning the model is still a problem, and this model is also missing portions of the arms due to limitations of scanning range. Using a cadaver provides challenges in itself, as body fluids tend to pool at the back of the body, organs shrink or swell, and airways collapse very soon after death. It has also been observed that the overall height of the body increases by several centimeters when it is lying (such as in an MR machine) as opposed to standing, for both live humans and cadavers. Software is often used to clean up and reposition these models and to change their size (adult to child, for instance) to ensure the most accurate electromagnetic modeling.

An additional problem with MR-scanned images is there is a trade-off between signal-to-noise ratio and a spatial shift that is seen between fat and water-based tissues such as muscle. When the signal-to-noise ratio is optimized, the fat will appear slightly shifted in location relative to muscle. The shift may be as much as 4 to 5 mm. In general this is a minor issue, as the majority of fat deposits are sufficiently large that this shift is inconsequential. While the fat shift may not cause much difficulty in defining the regions of fat, it does cause difficulty in defining the regions of skin. On the read side of the model, the fat obliterates the skin layer, making it appear very thin, while on the other side of the model, the skin appears very thick. A solution to this problem is to specify a predefined thickness of skin covering the whole body, and to apply this with a computer algorithm after image segmentation of the other tissues. This algorithm can be progressively refined as needed to control the thickness of skin throughout different regions of the body.

An additional consideration when developing a model for bioelectromagnetic simulations is the question of uniqueness of individuals. It has been shown that the height of a person affects how much current will be induced by high-voltage lines, and that the size of the head (children as compared to adults) affects the 1-g averaged SAR from cellular telephones. It has also been shown that minimal differences in 1-g averaged SARs from cellular telephones were obtained for several head models without the ear, and it is likely that differences in ear shape could affect the 1-g averaged SAR.

Once a tissue-segmented model has been developed, the electrical properties of the tissues are defined. The properties of human tissue change significantly with frequency, so it is essential to use data accurately measured at the frequency of interest. There is a wide range of published data on measured tissue properties, and work is still under way to measure and verify these properties. More information on the electrical properties of tissue can be found in Appendix A.

5.5 Energy Absorption (SAR)

This section discusses energy absorption and how it is controlled by the dimensions of the body and the wavelength of the fields. Specific absorption rate (SAR) represents power deposition in the body and is discussed in detail in Section 1.16. Simple spheroidal models are used to provide an intuitive understanding. These models can be solved analytically, which means they have a mathematical formula that describes the fields in the spheroid. If a person really were shaped like a spheroid, these models would be precisely accurate. The inaccuracy in these models comes from the inaccuracy in modeling, not the inaccuracy in the calculation. When more anatomically correct models are used to model whole-body effects instead of the spheroids, the precise values change, but the relative effects remain the same. For more localized effects (such as peak 1-g SAR calculations for cellular telephone dosimetry) the fields are highly dependent on the internal tissue distribution, and the anatomical models must be used.

As explained in Section 1.16, the penetration of incident EM fields into biological bodies decreases as frequency increases. The general effect is illustrated in Figure 1.40 by the strong decrease of skin depth as frequency increases. Because Figure 1.40 is for a plane-wave incident on a dielectric halfspace, though, it does not show other effects that are related to the size of the object. The energy absorbed by an object exposed to incident EM fields is a function of frequency, and of the size, shape, and electrical properties of the object. In this section, we discuss the characteristics of energy absorption in terms of the SAR (see Section 1.16), first at low frequencies where the wavelength is long compared to the object size, and then at higher frequencies where resonance effects can occur.

5.5.1 SARs at Low Frequencies

As explained in Section 2.2, the **E** and **B** fields are approximately uncoupled at low frequencies, that is, at frequencies low enough that the wavelength is large compared to the size of the object. In this frequency range, then, dosimetry consists of determining the internal fields due to the incident **E** field acting alone, and then due to the incident **B** field acting alone, as illustrated in Figure 2.1.

The nature of the internal fields at very low frequencies is illustrated by the calculation of the internal **E** fields in two very simple cases: a prolate spheroid exposed to an incident **E** field of 1 kV/m, which is typical of environmental fields, and a prolate spheroid exposed to an incident 60-Hz **B** field of 1 mT, which is typical of that of a hair dryer. These are the same as the examples illustrated in Figures 2.6 and 2.33. The prolate spheroid in each case is approximately the size of an average man. The conductivity, which dominates over the permittivity in determining the internal **E** at low frequencies, is 0.067 S/m, which is about two-thirds that of muscle tissue at 60 Hz.

One important consideration related to dosimetry is heating of tissue that might be caused by the internal **E** fields. To consider possible heating effects at low frequencies, let us first calculate from the SAR the temperature rise that would be produced by an internal peak **E** field of 1 V/m. Then we can calculate the temperature rise for any internal **E** field by multiplying our result by the square of that **E**, since the SAR varies as E^2. According to Equations 1.12 and 1.41, the SAR in the prolate spheroids for the above examples is given by SAR = σ_{eff} $E^2/2\rho$ = 0.067 × 1^2/2,000 = 0.0375 × 10^{-3} W/kg, if we assume the mass density of tissue is about the same as that of water, which is 1,000 kg/m^3. The power transferred to 1 g of tissue by the **E** would be 0.0375 × 10^{-3} × 10^{-3} = 0.0375 × 10^{-6} W. The 1 g of tissue would therefore absorb 0.0375 × 10^{-6} J of energy in 1 s, since energy is equal to power × time. Since 4.186 J are required to raise 1 g of tissue 1°C, the temperature rise of the tissue would be 0.0335 × 10^{-6}/4.186 = 8.96 × 10^{-9} °C, if all the absorbed energy were transformed into heat (and heat diffusion is ignored).

For the internal **E** fields in Figure 2.6, the temperature rise we calculated above for an internal **E** of 1 V/m would be multiplied by the square of the internal **E**. Thus, for the largest internal **E** of Figure 2.6 of 260 μV/m, the temperature rise would be 8.96 × 10^{-9} × (260 × 10^{-6})^2 = 6.06 × 10^{-16} °C. This is obviously a negligible rise in temperature. Even for the much larger internal **E** of 51 mV/m produced by the incident **B** of Figure 2.33, the temperature rise would be only 2.33 × 10^{-11} °C. Although these results are for a very simple model, they are consistent with the more general results that heating caused by typical low-frequency incident **E** and **B** fields is negligible.

Comparison of these calculated internal **E** fields with endogenous fields in biological systems also provides some perspective. Typical **E** fields across cell membranes are about 10^7 V/m. The threshold for nerve stimulation is typically about 6.2 V/m. Thus, internal fields of µV/m or even mV/m are small compared to typical endogenous fields. On the other hand, low-frequency incident **E** fields can produce shocks and burns. Because **E** fields at the surface could be much higher than internal **E** fields, low-frequency **E** fields might produce biological effects related to interactions at the surface.

5.5.2 SAR as a Function of Frequency

The general characteristics of the average whole-body SAR as a function of frequency for a model of an average man irradiated by an incident planewave with a power density of 1 mW/cm² (see Poynting vector in Section 3.4.3) are shown in Figure 5.5. The calculations shown in the figure are the results of early work using a combination of simple models (prolate spheroid, cylinder, capped cylinder), empirical techniques, and for part of the graph, interpolated estimations. Several different methods of calculation were used over various parts of the frequency range. At low frequencies, a long-wavelength approximation was made. The extended-boundary-condition method (EBCM) was used up to approximately the resonant frequency (see Section 3.6). Above that, the iterative EBCM (IEBCM) method was used. Other methods include the classical solution to Maxwell's equations for cylinders and the surface integral equation (SIE) method.

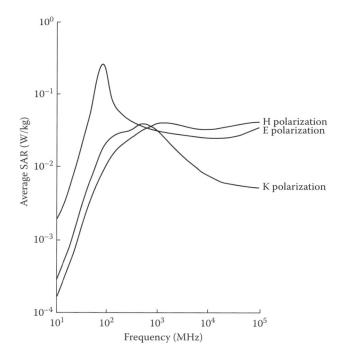

FIGURE 5.5

Average whole-body SAR as a function of frequency for models of an average man in free space for three polarizations, E, H, and K. The incident wave is a planewave with a power density of 1 mW/cm². (Adapted from Figure 3.39, Durney, C. H., et al., *Radiofrequency radiation dosimetry handbook*, 4th ed., Report USAFSAM-TR-85-73, USAF School of Aerospace Medicine, Aerospace Medical Division (AFSC), Brooks Air Force Base, TX, 1986.)

While more sophisticated methods can now be used because computing power has increased so much since these calculations were made, these early results are useful because they illustrate the elementary features of the SAR variation with frequency. Also, they illustrate the methods that were first used to understand the basic characteristics of SARs for humans and other animals. For all three polarizations, the SAR varies approximately as the square of the frequency at low frequencies. For E polarization, a resonance (see Section 3.6) occurs at about 80 MHz. Compared to the cavity resonance discussed in Section 3.6, the resonance shown in Figure 5.5 has a very low Q, which is due primarily to the much lower conductivity of the biological body. For long, thin metallic objects, like a wire antenna, a resonance occurs when the length of the object is about equal to a half wavelength. For biological objects, which are generally thicker and of lower conductivity, the resonance occurs when the object length is about four-tenths of a wavelength. More detailed simulations have since shown that the whole-body resonance is about 75 MHz for an ungrounded man model and about 37.5 MHz for a grounded man model (see Section 3.6). At frequencies above resonance, the SAR varies approximately as the inverse of the frequency.

At frequencies below resonance, the SAR for E polarization is highest, for H polarization is lowest, and for K polarization is in between the other two. Qualitative explanations of this effect are given in the next section.

To illustrate how the SAR frequency dependence varies with object size, Figure 5.6 shows the average whole-body SAR for a model of a medium-sized rat. The resonant frequency for E polarization in this case is about 600 MHz, considerably higher than for the average man. This is to be expected because the rat is much shorter than the man. Again, resonance occurs when the object is about four-tenths of a wavelength long. The resonance in the rat is less pronounced than in the man, probably because the man is relatively thinner than the rat. For the rat, the SAR also varies approximately as the frequency squared below resonance and as the inverse of the frequency above resonance.

Calculations in models with more realistic shapes show additional bumps on the average whole-body SAR graph that are caused by local resonance effects of the head, arms, and legs. The overall shape of the SAR graph is basically the same, however, as that obtained for simpler models. Inhomogeneous models have also been used to simulate the presence of organs and other details of the body. Calculations in inhomogeneous models show that the differing permittivities and conductivities of the various tissues of the body can cause the SAR to vary throughout the body. This local variation can also be a strong function of frequency. Even in homogeneous models, the SAR can vary significantly over the body, as illustrated in Section 3.4. In some cases, areas of intense local SARs can occur. These are sometimes called hot spots, which is not precise terminology because *hot* refers to temperature, while temperature inside the body depends not only on the absorbed energy but also on the thermal properties of the body.

5.5.3 Effects of Polarization on SAR

Figures 5.5 and 5.6 show that the SARs for both man-sized and rat-sized models vary significantly with polarization. Not only in these two cases, but also in general, polarization has a strong effect on SARs. This effect can be explained in terms of two general behaviors described in Sections 2.3 and 2.5 in connection with Figures 2.6 and 2.33. These general behaviors are: (1) the internal **E** field is generally greater when the $\mathbf{E}_{inc}$ is mostly parallel to the body surface than when it is mostly normal to the body surface, and (2) the internal **E** is generally greater when the cross-sectional area intercepted by the $\mathbf{H}_{inc}$ is greater than it is when the intercepted cross-sectional area is smaller. (Sometimes we use **B** and **H** almost

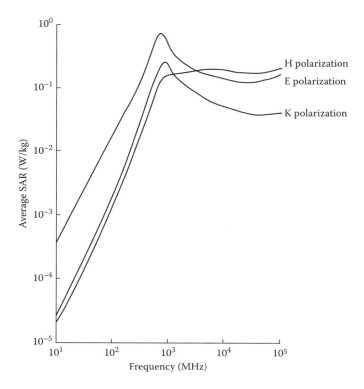

FIGURE 5.6

Average whole-body SAR as a function of frequency for models of a medium-sized rat in free space for three polarizations, E, H, and K. The incident wave is with a planewave power density of 1 mW/cm². (Adapted from Figure 3.40, Durney, C. H., et al., *Radiofrequency radiation dosimetry handbook*, 4th ed., Report USAFSAM-TR-85-73, USAF School of Aerospace Medicine, Aerospace Medical Division (AFSC), Brooks Air Force Base, TX, 1986.)

interchangeably, since $\mathbf{B} = \mu\mathbf{H}$ and for biological systems the permeability μ is approximately equal to μ_0, that of free space.)

These explanations are based on the low-frequency concepts that the $\mathbf{E}$ and $\mathbf{H}$ are approximately uncoupled at low frequencies, and that the internal $\mathbf{E}$ is the sum of the internal $\mathbf{E}$ produced by the $\mathbf{E}_{inc}$ and by the $\mathbf{H}_{inc}$. At higher frequencies, the effects are more complicated because $\mathbf{E}$ and $\mathbf{H}$ are coupled together and can strongly interact, but the ideas may have some validity even at higher frequencies. To facilitate the following discussion of how polarization affects SARs, let us define the internal $\mathbf{E}$ field produced by $\mathbf{E}_{inc}$ as $\mathbf{E}_{Eint}$ and the internal $\mathbf{E}$ field produced by $\mathbf{H}_{inc}$ as $\mathbf{E}_{Hint}$.

Now, from Figure 5.2 we see that for E polarization, $\mathbf{E}_{inc}$ is mostly parallel to the body and the cross-sectional area intercepted by $\mathbf{H}_{inc}$ is large compared to that of the other polarizations. Thus, for E polarization, both $\mathbf{E}_{Eint}$ and $\mathbf{E}_{Hint}$ are relatively strong. For H polarization, on the other hand, $\mathbf{E}_{inc}$ is mostly normal to the body, and the cross-sectional area intercepted by $\mathbf{H}_{inc}$ is smaller than that for E polarization. Therefore, both $\mathbf{E}_{Eint}$ and $\mathbf{E}_{Hint}$ are relatively weaker than for E polarization. The SAR for E polarization is thus greater than that for H polarization. For K polarization, E is mostly normal to the body, but the cross-sectional area intercepted by $\mathbf{H}_{inc}$ is large. $\mathbf{E}_{Eint}$ is therefore weak, but $\mathbf{E}_{Hint}$ is strong. The SAR for K polarization, therefore, is less than that for E polarization, but greater than that

TABLE 5.1

Summary of Explanations for the Effects of Polarization on SARs

Polarization	E_{inc}	H_{inc}	E_{Eint}	E_{Hint}	Relative SAR
E	Mostly parallel	Intercepts large cross section	Strong	Strong	Highest
K	Mostly normal	Intercepts large cross section	Weak	Strong	Middle
H	Mostly normal	Intercepts small cross section	Weak	Weak	Lowest

for H polarization, as shown in Figures 5.5 and 5.6 for frequencies below resonance. These ideas are summarized in Table 5.1.

5.5.4 Effects of Object Size on SAR

Because resonance for E polarization occurs when the object length is about four-tenths of a wavelength, it is obvious that the SAR is strongly dependent on object size. This dependence is further illustrated in Figure 5.7, which shows SARs for an average man and a medium-sized rat for E polarization, both plotted on the same set of axes. The whole-body

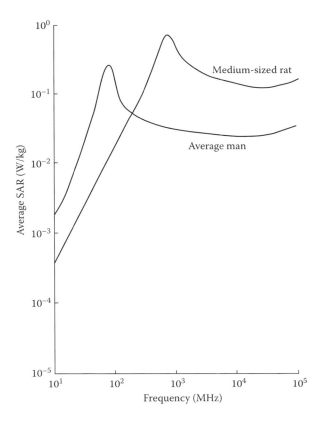

FIGURE 5.7
SARs from Figures 5.5 and 5.6 for E polarization for an average man and for a medium-sized rat plotted on the same set of axes.

average SAR that is plotted in the figures is obtained by calculating the total energy absorbed by the object per unit time and dividing by the mass of the object. The SAR, of course, varies from point to point inside the object. Because the rat is quite different in both size and anatomical features, the spatial distribution of the SAR inside the rat is quite different from that inside the man.

Even for objects of the same basic shape, but different in size, the internal SAR distribution can be quite different, depending on the frequency. For example, consider two prolate spheroids, one man-sized and one rat-sized. At 80 MHz, the internal SAR inside the man-sized spheroid would vary significantly over the volume of the spheroid. On the other hand, at 80 MHz, the internal SAR inside the rat-sized spheroid would be relatively constant. As first explained in Section 1.14, it is a matter of the size of the object compared to the wavelength. The internal SAR pattern is different in the two spheroids because the ratio of the object size to the wavelength is quite different.

On the other hand, the internal SAR distribution in the rat-sized spheroid at 600 MHz would be similar to the internal SAR distribution in the man-sized spheroid at 80 MHz, since the wavelength-to-object-size ratio is about the same in both cases. Thus, as explained in more detail in the next section, to get similar internal SAR distributions in two objects of different sizes, each object should be irradiated at a frequency for which the ratio of object size to wavelength is approximately the same.

The SAR in an object can also be affected by the presence of other objects. For example, when an object is placed on a perfectly conducting plane, the resonant frequency is approximately cut in half compared to the resonant frequency of the object in free space. Thus, for a man standing on a perfectly conducting plane and in good contact with the plane, the resonant frequency would be about 40 MHz (37.5 MHz according to Table 3.1). This effect occurs because the conducting plane has the effect of mirroring, or imaging, the man, making him appear to be twice as tall. For a man standing on the ground, which is not perfectly conducting, the resonant frequency would be lower than for free space, but not half as it would be for a perfectly conducting plane. Also, shoes would insulate the man from the ground, further affecting the resonant frequency.

5.6 Extrapolating from Experimental Animal Results to Those Expected in Humans

Because bioelectromagnetic experiments that require radiating people usually cannot be performed because of possible hazards, much research has been done with experimental animals to determine effects that might be expected in humans. The internal E fields in experimental animals exposed to specific EM fields usually differ significantly from those that would be induced in humans by the same EM fields because experimental animals differ significantly in size and shape from humans. But it is the internal EM fields, not the incident EM fields, that would cause any biological effects. Extrapolating results from experimental animals to those expected in humans must therefore be done with extreme care, and in some cases, such extrapolations may not be meaningful.

For example, studying effects on people exposed to the EM fields produced by cellular telephones by studying effects on rats exposed to the same EM fields is difficult. The ideal situation would be to produce the same internal E and H field distribution in the head of a rat as that produced in the head of a human using a cellular telephone. This is not possible, because the shape of the rat head is quite different from that of a human. Even to produce an EM pattern in the brain of a rat similar to that in a human brain at the same frequency is practically impossible because the brains are so different in size. The difference in the ratio of the size of the rat brain to the wavelength compared to the ratio of the size of a human brain to the wavelength would cause the internal EM fields to be very different in the two brains. If an effect in the human were caused by an internal **E** field with a given frequency and intensity acting on a given site in the brain, that would seem to be impossible to detect in experiments with rats.

On the other hand, it is possible to adjust conditions so that similarities in dosimetry between experimental animal exposure and human exposure are increased. In experiments where the biological effects are expected to be caused by heat resulting from absorbed EM energy, for example, frequencies of irradiation can be adjusted so that approximately the same whole-body SAR occurs in the experimental animals as would occur in humans. This same kind of adjustment would be appropriate for any experiment in which the biological effect is assumed to be caused by absorbed EM energy independently of the frequency.

To illustrate this kind of adjustment, we consider the following situation. If an average man were exposed to a 50-MHz E-polarized planewave with a power density of 1 mW/cm^2, what radiation would produce an approximately equivalent whole-body average SAR in a medium-sized rat? Let λ_r and λ_m be the wavelengths of radiation of the rat and man, respectively, and h_r and h_m be the heights of the rat and man, respectively. The first step is to find the f_r that would make

$$\lambda_r/h_r = \lambda_m/h_m \tag{5.1}$$

From Durney (1986, pp. 5.35–5.36), for prolate spheroidal models, $h_r = 0.2$ m and $h_m = 1.75$ m. From Equation 1.15, $\lambda_r = c/f_r$, where $c = v_p = 3 \times 10^8$ m/s, the velocity of propagation in free space. Similarly, $\lambda_m = c/f_m$. Substituting these relations for λ_r and λ_m into Equation 5.1 and solving for f_r gives

$$f_r = f_m h_m/h_r = (50 \times 10^6 \times 1.75)/0.2 = 438 \text{ MHz} \tag{5.2}$$

Thus, the rat should be irradiated at a frequency of about 438 MHz.

The next step is to find the incident power density at which the rat should be irradiated to produce approximately the same average whole-body SAR in the rat as in the man. From Figure 5.5 (note that the scales are both logarithmic), the average SAR in an average man at 50 MHz irradiated at 1 mW/cm^2 with E-polarization is about 0.09 W/kg (this can be read more easily from Figure 6.3 in Durney [1986]). From Figure 5.6, the average SAR in a medium-sized rat irradiated at 438 MHz with an incident power density of 1 mW/cm^2 is about 0.4 W/kg (read more easily from Figure 6.17 in Durney [1986]). To reduce the average SAR in the rat to 0.09 W/kg, the incident power density should be (0.09/0.4)(1) mW/cm^2 = 0.23 mW/cm^2. The rat, therefore, should be irradiated at a much higher frequency and lower incident power density to produce approximately the same whole-body average SAR as in the man.

We use this example merely to illustrate that dosimetry must carefully be taken into account in animal experiments in which the results are to be extrapolated to those expected in humans under similar conditions. The approximate equivalence in terms of average whole-body SARs will be appropriate in some experiments, but not in others.

5.7 Numerical Methods for Bioelectromagnetic Stimulation

Numerical computer simulations are often used to calculate the electric and magnetic fields in the human body. One of the main advantages of the numerical methods is their ability to handle complex geometries, in both the objects analyzed and the sources employed. Many important electromagnetic problems have configurations that are too irregular to be solved exactly with analytic methods. Numerical techniques break the complex configuration into small volume elements, or mathematical cells, each of which is assigned its own properties, such as conductivity and permittivity. The cells can also contain sources as needed. The electromagnetic equations are then solved for the collection of cells to obtain the fields in each cell. Various numerical techniques use different forms of the equations and employ different methods for their solution, but the goal of all is to find the fields in some sampled form throughout the volume of the model.

The shortcomings of the numerical methods revolve mostly around the ability to properly model the sources of interest and the portion of the body they impact. As electromagnetic applications in medicine become more and more focused and localized, the models must also become more detailed in that localized region. Models of the sources must also be more precise. The simulation methods available today are able to handle virtually all types of simulations of interest in bioelectromagnetics, and their progress is easily keeping stride with progress in application development. However, it is not uncommon that the level of detail available in the simulation will require huge computer resources (memory and computer time). In many cases, less detail will be utilized in deference to the amount of time we are willing to wait for the results. This is particularly true of imaging applications, where a major focus of imaging progress is obtaining sufficient accuracy with much less computation time. Also, simulations are not exact solutions, so it is always important to verify their accuracy, particularly for ranges of frequency and resolution, or for difference types of sources than have previously been verified. The exact analytical solutions for simpler cases (such as prolate spheroids) are often used for this verification. Measurements are another common method of validating complex bioelectromagnetic simulations, realizing of course that they have their own potential sources of error. Numerical simulations are powerful tools in the analysis of many electromagnetic problems and are used extensively in virtually all application areas.

Most often numerical methods are applied by using anatomical models such as those described in Section 2.6 to create block models of the body with resolutions on the order of millimeters. The model specifies what tissue type is present in each block of the model. The electrical properties of the tissues are then specified at the frequency of interest, as in Appendix A. When the sources of power are localized (such as for cardiac ablation or electrostimulation), the simulation can be simplified by using a model of only the part of the body that is impacted by the electromagnetic source. The type of simulation used depends mainly on the frequency range of interest, although the type of body model used and the precision with which the source must be modeled also help determine the best type of simulation to use. Methods commonly used for numerical simulations include the impedance method, finite-difference frequency-domain (FDFD) method, finite-difference time-domain (FDTD) method, and finite-element method (FEM). Combinations of these methods (hybrid methods) are also sometimes used.

Frequency-domain approaches assume the source is sinusoidal steady state. This means that a sine wave is assumed to have been propagating for a long time, and any transient effects have died away. Frequency-domain approaches calculate the magnitude and phase of the sinusoidal field at every point in the numerical model. Time-domain approaches use any form of the source (sinusoidal, pulsed, and so forth). The calculation begins when the field is initially turned on, thus modeling all of the transient behavior. If steady-state behavior is desired (which is typically the case for sinusoidal sources), the simulation is allowed to run until it reaches steady state.

5.7.1 The Finite-Difference Time-Domain (FDTD) Method

The finite-difference time-domain (FDTD) method is probably the most widely used computational method for bioelectromagnetic dosimetry because of its versatility and computational efficiency. It is particularly well suited to bioelectromagnetics because it can efficiently model the heterogeneity of the human body with high resolution (often on the order of 1 mm), can model anisotropy and frequency-dependent properties as needed, and can easily model a wide variety of sources coupled to the body over an extremely wide range of frequencies, from 60 Hz to 16 GHz, and also for broadband applications.

The FDTD method solves Maxwell's equations in the time domain (Equations 1.3 and 1.5) for the six vector components of the electric and magnetic fields in a model of the body made up of small cuboidal cells, such as the one shown in Figure 5.8. The tissue in each cell determines the electrical properties (ε, σ) that are used to calculate the fields in that cell as a function of time. The FDTD method effectively creates movies of how the fields penetrate and interact with the body, which can often provide excellent insight into the applications of interest. For instance, watching the field or current distribution on the implanted antennas described in Section 3.7 can guide their design. The major constraint for FDTD simulations is that the cell size (Δx) should be smaller than the smallest wavelength in the body divided by 10. The smallest wavelength is generally found in the tissue with the highest water content at the highest frequency of interest. Typically, steady state is reached after the wave has propagated through the body and back to the source one to three times, although this depends on the resonant nature of the body and sources. If the body is more resonant, the fields will reverberate longer within the body before reaching steady state. If the body is more lossy, the fields will be absorbed with fewer reverberations (less resonance).

Although originally applied only for mid-frequency problems, the FDTD method has been extended to both low- and high-frequency simulations. For low-frequency simulations, the body is so small compared to the wavelength that steady state is reached before

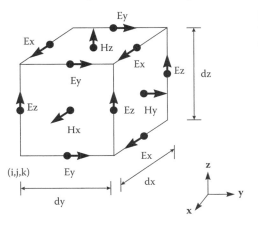

FIGURE 5.8
The FDTD cell showing distribution of the electric and magnetic field components.

a significant portion of the low-frequency wave has passed through the body. This origi-nally made it difficult to compute the steady-state magnitude and phase. Millions of time steps would have been required to do this in a direct fashion, and by then, the errors in the simulation would have added up to make the results unusable. A more efficient method of finding the steady-state magnitude and phase has solved this problem (see Section 5.7.1.1), and the FDTD method has now been applied to applications as low as 60 Hz.

For high-frequency applications, the model is very large compared to the wavelength of the source, or for very large mid-frequency applications, the model may be several wave-lengths or many cells long, making the models unreasonably large. For example, modeling a human holding a cell phone in an operating room (to determine the effect of cell phones on medical equipment) requires simulation of a millimeter-resolution human body (tens of millions of cells), as well as the operating room (billions of cells). The problem was solved by using FDTD to model the fields in and very near the human, and then using ray tracing of the waves to determine where they propagated and reflected within the operating room. This hybrid approach has extended the FDTD method to high-frequency applications where the model is very large. For simpler high-frequency applications, where the model is relatively small (optical applications), traditional FDTD approaches are sufficient.

The FDTD approach gives very accurate results if applied correctly. One such example is the modeling of a hertzian (infinitesimal) dipole at 900 MHz located 1.5 cm from a 20-cm-diameter brain-equivalent ($\varepsilon_r = 43.0$, $\sigma = 0.83$ S/m) sphere. The cubical cell size is $\Delta x = \Delta y = \Delta z = 5$ mm, which makes the sphere forty cells in diameter. Figure 5.9 shows the relative SAR along the y axis from the front edge of the sphere calculated using the FDTD method and compared to an analytical solution.

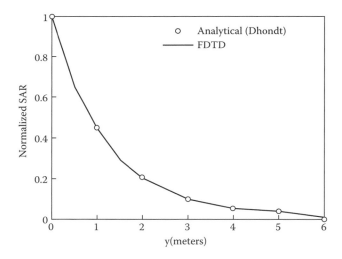

FIGURE 5.9
Relative SAR distribution along the y axis of a homogeneous brain-equivalent sphere excited by an infinitesi-mal dipole. (Analytical solution from Dhondt, G., and Martens, L., A canonical case with an analytical solution for the comparison of electromagnetic field solvers, paper presented at Proceedings of the COST 244 Meeting on Reference Models for Bioelectromagnetic Test of Mobile Communication Systems; FDTD from Furse, C. M., and Gandhi, O. P., Calculation of electric fields and currents induced in a mm-resolution human model with a novel time-to-frequency domain conversion, paper presented at IEEE Antennas and Propagation Society Conference, Baltimore, MD, July 21–25, 1996.)

5.7.1.1 Computation of Fields in a Human under a 60-Hz Power Line

Bioelectromagnetic simulations at low frequencies using the FDTD method require special consideration. For a 1-mm-resolution simulation of a man under a 60-Hz power line, for instance, over 10^{10} time iteration steps are needed to run a single cycle of the wave. Frequency scaling is often used at quasi-static frequencies where the **E** and **H** fields are assumed to be uncoupled. In that case, the simulation can be run at a slightly higher frequency (f′, still in the quasi-static range) than the actual frequency of interest (f), and the results can be linearly scaled to the lower frequency using

$$\mathbf{E}(f) = \frac{f}{f'}\mathbf{E}'(f').$$ (5.3)

The simulation is run using the tissue properties at frequency f, so that no scaling of the tissue properties is required. Suppose, for example, that the FDTD frequency f′ = 10 MHz is used, and scaled to f = 60 Hz. A single cycle (4,580 time steps) of the 10-MHz wave can then be used in the FDTD simulation. Also for low frequencies, the simulation generally converges in far less than a single cycle (because the body is so small compared to a wavelength), significantly reducing the overall computational requirement. A very similar approach was used to calculate the fields shown in Figure 2.39.

5.7.1.2 Computation of SAR from Cellular Telephones

The FDTD method has been applied to myriad mid-frequency simulations, including calculation of SARs and induced currents in the human body for a number of exposure conditions, exposure to the leakage fields of parallel-plate dielectric heaters, exposure to electromagnetic pulses (EMPs), exposure to annular phased arrays of aperture, dipole, and insulated antennas for hyperthermia, coupling of cellular telephone fields to the head, and exposure to radio frequency (RF) magnetic fields in magnetic resonance imaging (MRI) machines.

Determination of the SAR distribution in the head from cellular telephones and assessment of this with respect to the RF exposure guidelines is one of the best-known bioelectromagnetic applications of FDTD. Older-model cellular phones with the antenna protruding straight out of the body of the phone were found to deposit 40 to 50% of their radiated power in the head. The absorption was even larger for a child, as seen in Figure 5.10, because the antenna was physically closer to the child's head than the adult's. This significantly altered the radiation patterns from these phones and also the matching characteristics of the antenna.

Elements of cellular telephone simulations that have been found to significantly affect the accuracy of the simulation include the size of the metal case of the telephone and the dielectric properties used for the head. It has been shown that several different head models (with ears removed) can provide similar results, although homogeneous models have been found to significantly overestimate the 1-g averaged SAR value (by roughly 30%). Two of the most significant parameters affecting the power deposition in the head from the cellular telephone is the nature of the antenna (length, shape, and such) and how close it is to the head.

The tendency for electric fields to concentrate on metal corners (described in Section 2.4) shows up in simulations just as it does in physical structures. Remember that the numerical

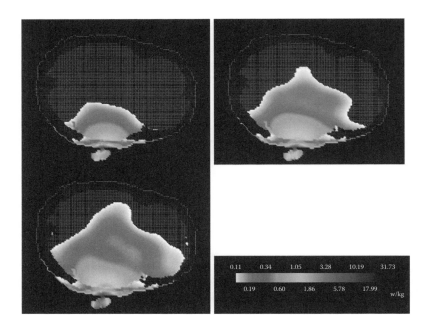

FIGURE 5.10
Please see color insert following page 146. Specific absorption rate computed for an 835-MHz cellular telephone. Top left, adult; top right, 10-year-old; bottom left, 5-year-old. (From Gandhi, O. P., et al., *IEEE Trans. MTT*, 44, 1884–97, 1996. © IEEE, 1996.)

simulation is modeling the objects by a set of cuboidal cells. If the phone is tipped, the metal antenna and metal components of the case also end up tipped. The square cells that model them have a stairstepped appearance on the edges, and thus have a series of sharp metal corners. These corners are not realistic, of course, but the modeling software calculates the fields as though the corners were actually present. The fields concentrate on these artificial corners and perturb the results. This effect is seen in all materials but is most pronounced in metals. A better way of modeling a telephone in a realistically tilted position is to maintain the upright telephone in the model and tilt the dielectric head instead. This artificial stairstepping effect should be kept in mind when modeling all metal objects with any simulation method. Coils, rods, plates, and other metal objects all are susceptible to this type of modeling error.

As an example of the effect of the many parameters impacting cellular telephone modeling, Table 5.2 shows a comparison of three different orientations of the head (shown in Figure 5.11) for a 2.76 × 5.73 × 15.5 cm telephone at 835 MHz, covered with 1 mm of plastic,

TABLE 5.2

Comparison of the 1-g Averaged SARs for a Cellular Telephone Next to the Head as a Function of Phone Position

Frequency (MHz)	Vertical Head Model	Tilted 30° Head Model	Tilted 30° Head Model with Additional Rotation of 9°
835	2.93 W/kg	2.44 W/kg	2.31 W/kg
1,900	1.11 W/kg	1.08 W/kg	1.20 W/kg

Source: Lazzi, G., and Gandhi, O. P., *IEEE Trans. Electromagnetic Compatibility*, 39, 55–61, 1997.

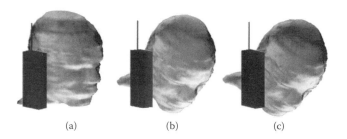

(a)   (b)   (c)

FIGURE 5.11
Please see color insert following page 146. Orientation of an 835-MHz cellular telephone near the head.

with a $\lambda/4$ antenna, also coated with plastic. The phone model is held against a model of the human head, and the simulation has an overall resolution of $1.974 \times 1.974 \times 3$ mm. Three values are shown, one for the phone held vertical to the head, touching the ear, which is pressed against the head. The second model has the phone tilted toward the mouth, but not pressed against the cheek, and the third model has the phone tilted toward the mouth and pressed against the cheek. As the phone is tilted toward the mouth, the antenna is effectively tilted away from the head, thus lowering the localized values very near the antenna, and consequently the 1-g averaged SAR value. This effect is most notable for physically long antennas. For the shorter antenna used at 1,900 MHz, the 1-g averaged SAR is not lowered significantly as the phone is tilted. This is because the antenna remains very near the head, despite being tilted.

5.7.2 The Impedance Method

At low frequencies, body tissues can be modeled as being purely resistive. The block model of the body then consists of a mesh of millions of resistors whose values depend on the tissue in each leg of the mesh. A voltage at one point on the body (let us say a potential caused by the heart) will create currents through all of the resistors to reach the ground point (perhaps an electrode on the lower leg). The currents in each individual resistor can be calculated using Ohm's law (see Section 1.8), and the power or SAR can be calculated from the voltages and currents distributed throughout the mesh. This method can be used for DC calculations as well as AC calculations at low enough frequency that the **E** and **H** fields can be assumed to be decoupled, typically lower than about 10 MHz.

The impedance method models the body tissues with their resistive, capacitive, and inductive components as shown in Figure 5.12. This figure shows one of the many millions of cells used to model the body. The cells are three-dimensional parallelepipeds, in general with a nonuniform rectangular cross section. Unique electric and magnetic field values are assigned to each face or each edge of the cells. Using the resistive, capacitive, or inductive cell, Maxwell's equations can be reformulated in terms of more familiar circuit equations. Thus, the technique is also known as the *equivalent-circuit* method. **E** fields are related to edge voltages, and **H** fields are related to loop currents.

The method then numerically solves the partial-differential form of the circuit equations by replacing the continuous derivatives with discrete differences; therefore, it is a finite-difference method. The solution assumes sinusoidal variation for all the current and voltage values, and finds steady-state sinusoidal answers. It is therefore a frequency-domain method.

The circuit equations written for all of the cells result in a complex, symmetric matrix equation. On the left side of the equation a very sparse matrix of circuit impedance values

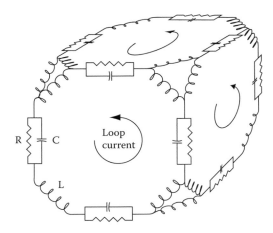

FIGURE 5.12
One of the mathematical cells employed in the impedance method. The model to be analyzed is divided into
numerous small cells like this one, each with its own
conductivity and permittivity. In general, the sides of
the cell are unequal in size. The impedance method is
also known as the equivalent-circuit method since the
circuit elements in each cell allow Maxwell's equations
to be solved by circuit analysis.

multiplies a vector of unknown loop currents; on the right side of the equation is a vector
of voltage sources. The solution for the unknown loop currents is obtained by an iterative
method, such as the Quasi-Minimal Residual (QMR) technique used in this study. Once
the steady-state loop currents are found, the corresponding **E** and **H** field values are read-
ily found for all cells throughout the model.

The chief disadvantage of the impedance technique is that it requires the solution of a
large matrix equation. Depending upon the size and robustness of the matrix, there can
sometimes be convergence difficulties in the iterative solution. Special measures (such as
prescaling of the matrix values) are needed to give good convergence of the solution.

5.7.2.1 Calculation of the E Fields Induced Near Implants During MRI

This section describes how the impedance method has been used to solve for the **E**
fields induced by the time-varying magnetic field of MRI in a patient who has metallic
implants.* There is not enough space to give full details of the technique or the results;
rather, we present this study as one example of the kinds of problems that can be handled
with numerical techniques, and we use a few of the results to illustrate concepts covered
earlier in the book.

The study described in this section was undertaken to determine whether the **E** fields
that are induced in a human body by the switched-gradient magnetic fields of MRI are
enough to cause nerve stimulation. This is an especially interesting question for patients
who have metal implants, because it might be expected that larger-than-normal **E** fields
would be induced in the tissue near metal implants by the magnetic field. It is also important to determine the pattern of these **E** fields, since, as has been seen several times earlier,
E fields tend to concentrate at the sharp edges of metallic objects.

In this study, patients are presumed to have an implanted spinal-fusion device consisting of a small DC generator with two thin metal wires running to some vertebrae in their
spine. The wires act as high-conductivity paths for currents induced in the body, altering
the **E** fields and current patterns inside the body.

---

* The results summarized in this section were obtained by Dale N. Buechler and can be found in more detail
  in his PhD dissertation, "A Finite-Difference Frequency-Domain (Equivalent Circuit) Method for Solving
  Maxwell's Equations," University of Utah, 1997.

5.7.2.2 Modeling an Implant in the Human Body

To represent the essential electromagnetic elements of the problem, we used a model containing $43 \times 59 \times 113$ cells, for a total of 286,681 cells. The cells were of variable size, ranging in dimension from 0.25 mm to 12.7 mm. Figure 5.13 is a view of the model with its various elements. Overall, the model was 11.6 cm wide × 13.5 cm deep × 75.9 cm long, about one-half the size of the thorax in cross section, but much longer. Inside, the lengths of the two conducting wires of the implant were modeled as well as a portion of the spine near the wires. A 0.67-cm-thick layer of fat was located at the border of the model nearest the spine. The remainder of the interior of the model was filled with a homogenous medium with the same properties as muscle.

As explained above, the metallic wires of the implant and the region around them were of the most interest, so they were modeled with the highest resolution possible using the smallest $0.25 \times 0.25 \times 0.25$ mm cells; smaller cells would have required too much computer memory. The two wires were connected in a U shape, as shown in the view of Figure 5.14. The wires were covered with a thin (0.25 mm) insulation layer over most of their length, except for the last 3.67 cm, which was left bare. A special case was also treated in which the wires were disconnected from the bottom of the U, leaving the two side wires alone and isolated (as if cut off or broken).

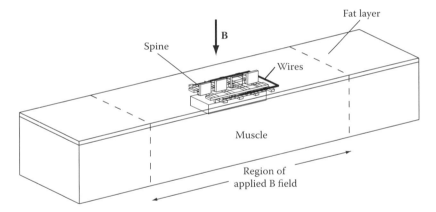

FIGURE 5.13
Overall model of wire implants near the spine. The **B** field is representative of the time-varying magnetic field encountered in MRI. It is uniformly applied over the central region of the model.

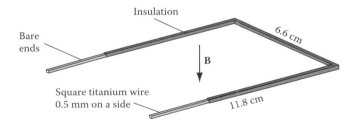

FIGURE 5.14
Dimensions of the wire implant. The wire is insulated (by a covering that has low conductivity) over much of its length, but is bare for a distance of 3.67 cm at both ends. The wire itself is assumed to have the conductivity of titanium. The **B** field is perpendicular to the plane of the U-shaped wire

E fields were calculated for two orientations of the magnetic field. (**B** was used here rather than **H**; for the nonmagnetic body, they are related simply by $\mathbf{B} = \mu_0\mathbf{H}$.) The orientation for which the implant had the highest impact on the field pattern was the one where the **B** field vector was perpendicular to the plane of the U-shaped wire, as indicated in Figure 5.14. This occurs because the wire shape in that orientation presents the largest cross-sectional area for intercepting the changing **B** field, as discussed in Section 2.6. The **B** field was assumed to be sinusoidal at a frequency of 600 Hz.

5.7.2.3 Results of the Numerical Calculations

All of the steady-state results were transformed into plots of the **E** field or current density vectors in various planes of interest inside the model. Calculations were first made without the wires present to establish a baseline for the **E** fields in the absence of the implant. Those results are shown in Figure 5.15 for a region near the spine. As expected, the **E** field lines (and resulting eddy-current paths) circle around the **B** field vectors, which are perpendicular to the figure. There is an elongated center of circulation for the **E** field lines near the center of the body. The **E** field magnitude is zero there, and increases toward the periphery of the body. This behavior is typical of the eddy patterns resulting from magnetic sources, and is very analogous to that seen in Figure 2.41 for a culture dish, and in Figure 6.3 for an H field hyperthermia applicator.

Then several cases were studied that included various configurations of the implanted wires; only two will be discussed here. The first case is for the complete U-shaped wire model (as shown in Figure 5.14) with a perpendicular **B** field. Figure 5.16 shows the resulting **E** field vectors in the plane containing the wire. The **E** field magnitudes are larger than in the absence of the wire implant, so they are plotted to a much different scale in this figure compared to the previous figure. It is obvious that the **E** fields are modified near the bare ends of the wire. This is due to the conducting wire in the presence of the

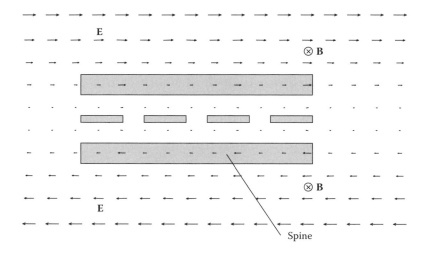

FIGURE 5.15
E field pattern induced in the model without the wire present. The eddy currents follow this same pattern since current density is proportional to E for a homogeneous medium (muscle in this model). These results represent a reference for comparison with the results when the implant is in place. The vectors shown are interpolated from a finer grid.

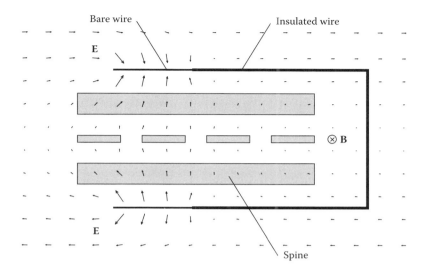

FIGURE 5.16
E field pattern with the U-shaped wire implant in place. Current is induced in the wire by the time-varying **B** field. The current exits one end of the wire and enters the other, causing enhanced **E** fields in the tissue near the bare wire ends. The vectors shown are interpolated from a finer grid.

time-varying magnetic field. Current is induced to flow in the wire, much like a secondary winding in a transformer. At the low frequency of this problem (600 Hz), the current must form a complete path and close on itself. Since the wire does not form a complete loop, the current flows out of the bare wire at one end of the open loop, passes through the tissue in between, then flows back into the bare wire at the other end, completing the closed path. It must flow out the bare ends since the rest of the wire is insulated.

The **E** fields in the tissue near the bare ends of the wire are enhanced due to this current flow. This is shown in close-up view in Figure 5.17. The **E** fields are larger (more concentrated) at the small end of the wire than they are at the sides of the wire, again illustrating the concept that **E** fields concentrate around sharp boundaries of conducting objects. The magnitude of the largest **E** field in this case is about ninety times greater than at the same location without the wires present (Figure 5.15).

The current pattern inside the wire U is also interesting to study. This pattern is shown in Figure 5.18. As explained above, most of the current induced in the partial wire loop flows out the open ends through the tissue to complete a closed path. But as Figure 5.18 shows, there is a *local* loop current that flows in a path contained inside the wire itself. This is evident by the unequal current density on the inside and outside edges of the wire. This local eddy-current pattern inside the wire is due to the finite cross section that the wire presents to the perpendicular **B** field. If the wire were infinitely thin, there would be no local circulating current within the wire itself, just the global circulating current that goes through the tissue.

The second case studied was with the bottom segment of the U-shaped loop removed, leaving only the two isolated side wires. The overall **E** field pattern is shown in Figure 5.19. The main difference here compared to the previous case is that since the partial wire loop is now broken, there is less of a complete conducting path for the current induced in the wire. It must now flow out through tissue at both ends of both wires. This reduces the magnitude of the induced current, and correspondingly reduces the **E** field magnitudes at

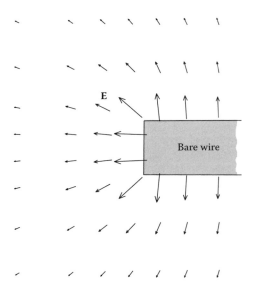

FIGURE 5.17
Close-up view of the E fields in Figure 5.16 near the wire ends. The fields are concentrated near the sharp corners of the wire.

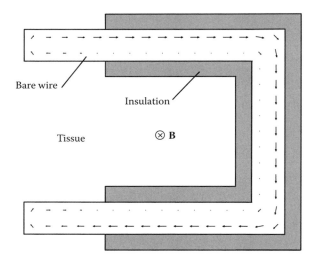

FIGURE 5.18
Current inside the wire. In addition to the global current loop indicated in Figure 5.16, there is a local loop of current inside the wire itself due to the finite cross section of the wire, which intercepts the perpendicular **B** field. The wire shape is not drawn to scale.

the bare ends of the wires. In this case, the maximum **E** field magnitude at the bare end of the wires is about seventeen times that at the same location without the wires.

However, at the other end of the wires, the **E** field is even larger. In fact, it is the largest value seen in any of the cases studied, about 197 times the magnitude at the same location without the wires being present. This is due entirely to the configuration of the insulation at this "broken" end of the wires. As shown in Figure 5.20, it is assumed that the insulation

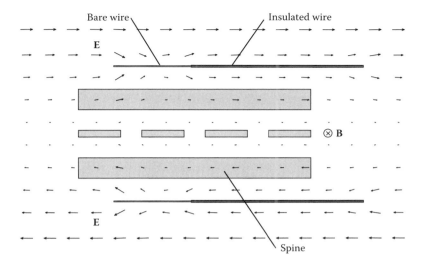

FIGURE 5.19
E field pattern when the two side wires are disconnected from the bottom cross wire and isolated. Less current flows in the side wires, and the global current loop must now flow through tissue at both ends of the wires. The vectors shown are interpolated from a finer grid.

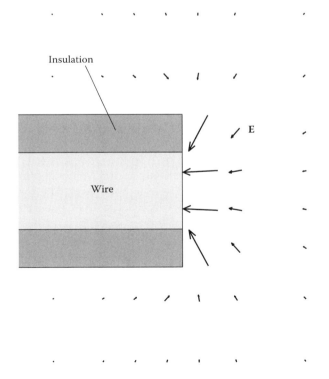

FIGURE 5.20
Close-up view of the **E** fields in Figure 5.19 near the broken end of each wire that is covered (except for the face) with insulation. Since the current in the wire must all flow through the small face, the current density is high here and the **E** fields are the largest for any case studied.

extends all the way to the end of the wires at this end, and only the face of the wire is uninsulated and exposed to the tissue. In other words, at this end the insulation is flush with the wires. Therefore, the current that is induced in the wires—even though smaller than before—is forced to flow entirely out the small wire ends, concentrating the current density and the **E** fields at these points. As indicated by the figure, however, the high values of **E** occur over a very small volume. Even for this worst case, it was determined that, for the assumed parameters of the problem, the **E** fields were not large enough to stimulate the nerves.

5.8 Electromagnetic Regulations

Dosimetry is used to determine the amount of power, fields, and current occurring in various parts of the body from different field exposures. Determining if these are safe or not requires an understanding of the biological effects of these fields, as described in Sections 1.18 and 6.5. These effects have been used to define allowable guidelines or regulations for electromagnetic field exposures. In addition to the allowable exposure strengths, methods of calculating or measuring the exposures are also generally specified in the guidelines, as well as allowed frequencies.

5.8.1 Allowable Frequencies

A number of government regulations impact the design of electromagnetic devices. The first is the allowable frequency. Applications that are used external to the body or for short periods of time (hyperthermia treatment, pain control, and cardiac ablation, for example) utilize the Industrial Scientific Medical (ISM) bands (433, 915, and 2450 MHz) in both the United States and Europe. Higher frequencies have the advantage of smaller antenna or applicator sizes, but the disadvantage of lower depths of power penetration within the body. Implantable medical devices that are meant to stay in the body for a long period have been allocated a band of their own in the United States, the Medical Implant Communication Services (MICS) band from 402 to 405 MHz. The maximum bandwidth that can be used by a single device is 300 kHz in this band. The maximum power limit is 25 μW equivalent radiated power (ERP). MICS shares its frequency allocation with the Meteorological Aids Service (METAIDS), which is used primarily by weather balloons, and it is therefore specified for indoor use.

5.8.2 Limits on Absorbed Power

The second type of bioelectromagnetic regulation is the limit on allowable absorbed power in the body. Limits on whole-body absorbed power (as defined by SAR) were established by the Institute of Electrical and Electronics Engineers (IEEE C95.1) in 1991 and adopted by the American National Standards Institute (ANSI) in 1992. Neither of these groups has regulatory authority over bioelectromagnetic applications; however, these standards have often served as the basis for regulations (with some modifications specific to certain applications and frequencies) from the U.S. Food and Drug Administration (FDA) and Federal Communications Commission (FCC) as well as their European and worldwide counterparts. The IEEE-ANSI standard covers frequencies from 3 kHz to 300 GHz.

The rationale for the IEEE-ANSI C95.1 standard is based on the thermal effects from power deposition in the body and the body's ability to dissipate this heat. A whole-body-averaged SAR of 0.4 W/kg for occupational situations and 0.08 W/kg for the general public was determined to be the maximum recommended power deposition in the body. This is derived from a 4-W/kg value that was found to result in reduced activity in laboratory animals such as rats and monkeys. This also corresponds to a human body metabolic rate of 4 W/kg for moderate activity such as light housecleaning. A safety factor of 10 was used to reduce this power level to the 0.4 W/kg power deposition level that is still used today.

The first standard for electric field exposure (prior to 1982) was based on a very simple spherical model. An incident power density of 10 mW/cm² was found to produce an SAR of 0.4 W/kg and was therefore chosen as the basis for the first bioelectromagnetic exposure standard. In 1982, the spheroidal models described in Section 5.5 showed that the human body (or major parts of the body) resonates at frequencies from about 25 to 225 MHz, depending on the size of the body, its orientation, and whether or not it is grounded. When in resonance, the body absorbs up to ten times as much power as when it is not in resonance. Thus, the exposure maximum was lowered to 1 mW/cm² near that frequency range.

Progressively more advanced dosimetric calculations were used in the development of the IEEE-ANSI C95.1 standard, which is given in terms of maximum permissible exposure limits, or MPE, shown in Figure 5.21. Over the frequency range from 3 kHz to 300 MHz, the limits are given in terms of maximum electric and magnetic field strengths, and over the range from 100 MHz to 300 GHz in terms of maximum power density. In the over-lapping region (from 100 MHz to 300 MHz) in Figure 5.21, the correspondence between power density levels and the respective electric and magnetic field strengths can be calculated from relationships found in earlier sections. Assuming that the electromagnetic field exposure can be modeled as a planewave, the Poynting vector gives the power density carried by the wave (Equations 3.9 and 3.10) as a function of either electric field or magnetic field strength. For exposure in free space, $\varepsilon = \varepsilon_0$, $\sigma_{eff} = 0$, and $\mu = \mu_0$. Using Equation 3.1 for the definition of wave impedance Z, it can be shown that $Z = 377\ \Omega$ in free space, and Equations 3.9 and 3.10 simplify to

$$P = \frac{E_{rms}^2}{Z} = \frac{E_{rms}^2}{377}\ (W/m^2), \tag{5.4}$$

and

$$P = Z H_{rms}^2 = 377\ H_{rms}^2\ (W/m^2). \tag{5.5}$$

Using these equations, the 1.0 mW/cm² limit (equivalent to 10 W/m²) between 100 and 300 MHz for controlled exposure then corresponds to an electric field strength of 61.4 V/m rms and a magnetic field strength of 0.163 A/m rms, as shown in Figure 5.21. Similar correspondence between power density and field strengths applies to the uncontrolled exposure limits. Also, based on anatomically-based models, it was determined that the magnetic field couples less efficiently to the body than does the electric field, so exposure guidelines are less stringent for magnetic fields at frequencies below about 3 MHz.

The IEEE-ANSI C95.1 standard recognizes two different classes of people. Controlled environments are those where the person understands the potential for exposure, and this is usually limited to occupational exposures. Uncontrolled environments are for the general public, where a person has no knowledge or control of his or her exposure. In the resonance region, the uncontrolled exposure is reduced by a factor of 5 from the controlled exposure. The factor of 5 comes from a precedent set by the Massachusetts Public Health Department, which adopted the two-tier policy with a factor of 5 reduction (based on the 1982 standard) for uncontrolled exposures in 1983.

5.8.3 Localized Exposure Limits

Prior to the 1980s, there were very few, if any, localized exposure applications of bioelectromagnetics. The 1980s brought questions about the safety of hyperthermia applicators, localized sources from hair dryers, electric drills, and other handheld devices, and handheld radio phones. Thus, a guideline was needed for localized exposure in addition to whole-body averaged exposure. The IEEE-ANSI C95.1 guideline provides that the localized SAR in any 1 g of tissue may be relaxed to 8 W/kg (from the original 0.4 W/kg) for controlled exposures. In extremities of the body (hands, wrists, feet), this was relaxed even further so that a 10-g SAR can be 20 W/kg for controlled exposures. For uncontrolled exposures (the general public), this localized guideline was reduced by a factor of 5, leading to the well-known 1-g SAR exposure limit of 1.6 W/kg. This guideline is applied most often in the certification of cellular telephones by the U.S. FCC, which adopted this standard. Other countries have followed variants of the IEEE-ANSI guideline.

5.8.4 Induced Current and Shock Guidelines

Another aspect of the IEEE-ANSI C95.1 standard is for induced current and electric shock. The plateaus on the left of Figure 5.21 cap the electric and magnetic fields to reduce the possibility of spark discharge for most exposure conditions. This limit prevents most, but not all, spark discharge conditions. Spark discharge depends on many factors, including the dryness and condition of the skin, the size of the person, whether or not he or she is grounded, and the area of the body making contact with the object. Electromagnetic spark discharge is similar to the static spark that occurs when charges collect on a person (perhaps

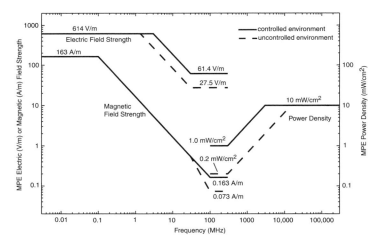

FIGURE 5.21
Maximum permissible exposure (MPE) limits for electric and magnetic fields and power density according to IEEE C95.1 (1991).

TABLE 5.3

Maximum Induced Current Limits from
IEEE-ANSI C95.1

	f = 0.003–0.1 MHz	f = 0.1–100 MHz
Controlled Environment		
Both feet	2000 f	200
Each foot	1000 f	100
Contact	1000 f	100
Uncontrolled Environment		
Both feet	900 f	90
Each foot	450 f	45
Contact	450 f	45

Note: Values are given in mA rms, with f in MHz. The averaging time is 1 s. Foot limits assume a free-standing individual with no contacts. Contact limits are for touching a metal object.

wearing nylon socks) and are discharged through a spark when that person touches a grounded metal object such as a light switch. Electromagnetic fields charge up an object that is insulated from the ground, such as a car or a bus. The object acts like a capacitor to store this charge. An unsuspecting person who is grounded or nearly so touches the object and receives a spark discharge. A car or bus requires about 1.1 to 3.3 kV/m electric field in order to charge it sufficiently to create a painful spark discharge to a person standing on a ground plane. The IEEE-ANSI C95.1 guideline would prevent this spark discharge. But for a metal fence wire that is on insulated posts, only 500 V/m is required, and this would not be prevented under the C95.1 standard.

Furthermore, at high frequencies, sparks can be discharged with much lower field values. One example is on an aircraft carrier where RF communications antennas are placed on the metal deck near the large metal airplanes. A voltage can be generated on the plane that is sufficient to produce burns when someone touches the aircraft. When the communications frequency is high enough that the wavelength is small compared to the size of the aircraft, many different hot spots can exist on the aircraft. These hot spots are somewhat unpredictable, as they depend on the orientation of the aircraft relative to the antenna. The U.S. Navy uses a voltage rather than electric field standard. This is easier to measure and makes more sense for point contact measurements. The maximum allowed voltage is 140 V rms. Because of the highly variable nature of the potential on the surface of the plane, this may still not be sufficient to eliminate all burns, but it is certainly more stringent than the C95.1 guideline in this respect.

The induced current limits in the C95.1 standard are based on the threshold of perception when a person is standing on a metal ground plane, grasping a rod electrode in the palm of his or her hand. Table 5.3 gives the maximum induced current limits from IEEE-ANSI C95.1.

5.8.5 Power-Line and Static Field Limits

The IEEE-ANSI C95.1 guidelines do not cover 60-Hz (U.S.) or 50-Hz (elsewhere) power frequency fields. The power line frequencies are so low that the wavelength is very large compared to the body size, and heating (which is the basis of the C95.1 guidelines) does not occur in the same way. Bioeffects at power-line frequencies remain hotly debated, and the wide variance in the guidelines mirrors this debate. Guidelines have been proposed by the European Agency for Electrotechnical Standards (CENELEC), the National Radiation Protection Board in the UK (NRPB), the American Conference on Governmental Industrial Hygienists in the United States (ACGIH), and the International Committee on Non-Ionizing Radiation Protection (ICNIRP). Also, the IEEE C95.6 (2002) guidelines cover the frequency range 0 to 3 kHz. Some individual states and countries have adopted power-frequency standards. Power-frequency standards have large variation, with magnetic field limits ranging from 0.2 µT to 21 mT and electric field limits ranging from 10 V/m to 30 kV/m.

Static exposure guidelines also vary widely. Perhaps the most salient to this book is the MRI exposure guideline. The International Electrotechnical Commission (IEC 2002) allows up to 2T for normal operating mode, 2T to 4T for first-level controlled operating mode, and above 4T for second-level controlled operating mode. The NRPB allows 2.5T for the head and trunk (4T under controlled conditions) and 4T for limbs. The International Radiation Protection Agency (IRPA) allows 2T for the head and torso and 5T for the limbs. Newer switched-gradient MRI machines present some concerns with regard to peripheral nerve stimulation (pain), which depends on the size of the individual and also on the placement and nature of the receiving coils.

Many of the bioelectromagnetic applications described in this book do not fall within the standard exposure guidelines. However, most guidelines make exception for therapeutic devices used under the direction of a physician, and the burden falls on the manufacturer to verify that the device is safe for use.

5.9 Conclusion and Summary

Dosimetry is a critical aspect of bioelectromagnetics. Chapters 2 to 4 describe several qualitative effects observed for these fields and some methods of calculating their quantitative values for very simple configurations. More advanced analytical methods exist to allow detailed calculation of other configurations. These analytical solutions are very valuable for providing physical insight and also for testing detailed dosimetric calculations. Numerical methods such as the finite-difference time-domain (FDTD) method, impedance method, and finite-element method (FEM) are most often used today to determine the fields, currents, and power in and around the body from realistic electromagnetic sources. These methods use detailed models of the human body, most often derived from MRI scans, and also detailed models of the electromagnetic sources. The fields are then computed with great resolution within the body or part of the body being considered. The dosimetric values can then be used to design electromagnetic devices and evaluate their induced fields to see if they fall within the allowable exposure guidelines. Virtually all bioelectromagnetic devices in the mid-frequency range require this type of detailed analysis. In fact, calculations are often substituted for internal field measurements, because they are simpler, less invasive, and equally or sometimes even more accurate. Fields from lower- and higher-frequency devices can often be measured rather than calculated, so numerical techniques are used less often in these ranges.

Dosimetric calculations also hold a key to future bioelectromagnetic devices because they allow extensive detailed understanding of the field interactions with the very complex structures in the human body. For instance, the new highly localized head, neck, and breast coils used for MRI described in Chapter 6 are designed with detailed numerical modeling programs (mostly FDTD and FEM). Miniaturized NMR spectroscopy devices are also evaluated in detail using FEM.

Our understanding of the interaction of cellular telephones, power lines, and myriad consumer electrical devices with the human body has been significantly improved because of dosimetric evaluations. Dosimetry provides a unique look into the body, often in the early design stages before a device is even built. Another major use of dosimetry is to provide a better understanding of experimental setups and how the exposure setups affect the exposure conditions, and hence, potentially, the biological responses.

References

Durney, C. H., Massoudi, H., and Iskander, M. F. 1986. *Radiofrequency radiation dosimetry handbook*, 4th ed. Report USAFSAM-TR-85-73, USAF School of Aerospace Medicine, Aerospace Medical Division (AFSC), Brooks Air Force Base, TX.

6

Electromagnetics in Medicine: Today and Tomorrow

6.1 Introduction

The previous chapters have described the basic ways that electromagnetic (EM) waves interact with the body and some of the applications that demonstrate those specific principles. This chapter will discuss the overall potential and challenges for bioelectromagnetics in medicine and provide a brief survey of the major technologies that are in use today and how they utilize the basic principles from the previous chapters. At the end of the chapter we will discuss the emerging technologies that are pushing the forefront of bioelectromagnetics.

Bioelectromagnetics for medicine spans applications in imaging, communication with implantable medical devices, heating for treatment of cancer, cardiac abnormalities, hypothermia, measurement of fields for assessment of radio frequency (RF) safety, augmentation of healing, reduction of pain, and electrical stimulation for cardiac and other therapy. Some of these applications have gained worldwide acceptance and are currently used with human subjects, and others are still in the research and development stage.

6.2 Fundamental Potential and Challenges

Electromagnetic fields have strong potential for medical applications because they can be transmitted, guided, and focused noninvasively or semi-invasively. They can transmit power (to be used for heating) or data (to be used for communication). They can induce or collect biological responses used for imaging, spectroscopy, biofeedback, EMG, and EKG, for example. Both the present and the future applications of bioelectromagnetics capitalize on the ability of electromagnetic waves to transmit or receive without directly contacting the point of interest. This ability to impact or be impacted by their remote environment is probably the greatest strength of electromagnetic fields. But it is also their greatest challenge. Not only does the environment of interest impact the electromagnetic field, but almost everything else nearby does also! The ability to receive signals from a single neuron, for instance, is highly desired, and it can indeed be received. But the signal from all of its neighbors is received as well. The ability to guide power into the body is great. But the structures within the body that are in the path of this guided signal change the field as it passes near or through them, altering the power in ways we did not intend. Our ability to harness the power of bioelectromagnetics is limited by how well we are able to control the propagation of the waves into and out of the body. As we will see in the final section of this chapter, methods to become more localized in this control are opening up the newest bioelectromagnetic frontiers.

Bioelectromagnetic applications are, of course, strongly impacted by the lossy dielectric materials that make up the human body. At low frequencies, the conductivity of the tissue dominates the behavior of the field, and at high frequencies, the dielectric values tend to dominate. Muscle and other high-water-content tissues are highly conductive and therefore very lossy. Fat and other low-water-content tissues have lower conductivity and therefore lower loss. These two tissues are near extremes in the body. Biological tissues absorb power (which is good if you are trying to heat that part of the body, but bad if you are trying to transmit through to another section of the body or out of the body for telemetry purposes). The absorption produces heat, which is the basis of major portions of the RF safety guidelines, and must be minimized in order to stay within them. The tissues also reflect power, which is good if you are trying to shield a section of the body, and bad if you are trying to transmit energy into the body. The standing waves produced as a result of this reflection also create the potential for otherwise unexpected overheating. If controlled in laboratory equipment, these same standing waves produce excellent, well-controlled sources for bioelectromagnetic research. Problems due to reflection are most prevalent at interfaces between high- and low-water-content tissues or between tissue and air. Let us review each of these effects in more detail, now that you have seen how they are used throughout the low/medium/high frequency ranges in Chapters 2 to 4.

The first characteristic is the lossiness of the body. This controls the penetration of EM fields into an object, which generally decreases with increasing frequency, as illustrated by Figure 1.40. This figure shows skin depth as a function of frequency for a planewave incident on a dielectric halfspace having the approximate electrical properties of the human body. At higher frequencies, the EM fields do not penetrate to any appreciable depth. For low-frequency applications, this means that the fields can penetrate well into the body, but with little control (they spread out widely). Focusing these low-frequency fields requires metallic applicators (electrodes) that are small and highly localized. At low frequencies, the lossiness of the body acts most like a resistive network and is often modeled as such. The mid-frequency range is the best range to consider for both penetration and some focusing. Thus, it has been chosen for applications such as biomedical telemetry (communication) and hyperthermia (heating for treatment of cancer). At very high frequencies, the fields simply do not penetrate into or out of the body. Thus, attempting to produce high-frequency internal EM fields in the interior of the human body by external applicators for any useful purpose is futile; a localized field requires localized applicators. In this range, optical applications abound for medical equipment. Lasers, lenses (often coupled with fiber optic cables), and the use of infrared measurement are some of the highlights of this frequency band.

The second characteristic of importance to bioelectromagnetic applications is the ability to guide, focus, or otherwise control the electromagnetic energy. The future of bioelectromagnetics is in the ability to more locally control or receive fields within the body. Traditional methods of controlling electromagnetic power generally can focus or concentrate the power to a region of size on the order of a wavelength. At low frequencies, therefore, where the wavelengths are long, EM energy cannot be tightly focused, except by localized, direct-contact, or near-direct-contact applicators (electrodes and the like). As an extreme example, consider EM fields at 60 Hz, for which the wavelength in free space, according to Equation 1.15, is 5×10^6 m. EM fields at 60 Hz thus cannot be focused to a region in free space smaller than about 5×10^6 m in diameter. In the human body, the wavelength inside the body at 60 Hz is nearly 10^4 times smaller than in free space because the relative permittivity of the body at this frequency is in the neighborhood of 10^8, as indicated by Figure A.1, and the permittivity slows the wave down in a way similar to that for spherical waves and planewaves (see Equation 3.4). In a medium having the electrical properties of the body, therefore, the EM fields at 60 Hz cannot be focused to a region smaller than about 500 m. Even at 1 MHz, the EM fields in a medium with the electrical properties of the body cannot be focused to a region smaller than about 9 m. This explains why electrodes are used to control the power at low frequencies: Electrodes can be placed directly where the power is desired, thus concentrating the power at their specific location.

The fundamental difficulty encountered in trying to produce EM fields inside the human body by external sources for useful medical purposes, then, is this: at high frequencies, the fields will not penetrate far enough into the interior of the body to be useful. Methods based on constructive or destructive interference, for example, which would work at high frequencies in relatively lossless objects, will not work in the human body because of the high attenuation. On the other hand, at low frequencies, the internal fields penetrate to the interior of the body, but they cannot be focused or localized because the wavelengths are so long compared to the size of the body. Furthermore, although the low-frequency internal fields will extend into the interior of the body, they are generally weak. In addition, the spatial distribution of the internal fields produced by external applicators is difficult to control. Consequently, producing and controlling desired EM fields in the interior of the human body is difficult at any frequency. The trade-offs between penetration and focusing, low and high frequencies, and overall challenges of controlling power within the body will become very clear when we explore hyperthermia for cancer therapy in Section 6.3. The bottom line is that power can be deposited in large regions of the body, and can even be focused on smaller regions. But if highly localized power deposition is required, once again, very localized applicators are used. Intriguingly, this concept of more localized applicators is also having a major impact on the next generation of magnetic resonance imaging (MRI) medical imaging devices.

Thinking again about the idea that our ability to harness the power of bioelectromagnetics is controlled by our ability to transmit or receive fields from a highly localized region, we need to think beyond just physical waves and our ability to focus and control them. MRI, described in detail in Section 6.4, typically uses a very powerful electromagnet to produce a strong, uniform magnetic field longitudinally (head to toe) in the body. This magnetic field causes the magnetic spin vectors of the nuclei in the hydrogen atoms in the body to align with the field, much like the needle on a compass. Superimposed on this DC magnetic field is a spatially varying magnetic field produced by several gradient coils. After a short RF pulse tips the spins, they relax (precess) back to their original orientation, and in doing so they induce an RF electromagnetic signal in external receiving coils. Traditional receiving coils are large enough to surround the whole body and receive the electric signals generated by all of these atoms returning to their natural state. The external coils have no way to focus the signals to a small region of the body, yet most MRI scans today have resolution on the order of 1 mm. Clearly this did not come from focusing of the fields. Instead, the resolution in MRI comes from using the gradient magnetic field across the body to adjust the frequency of the returned signal, and thus create a spatial code that tells where each portion of the returned signal came from. This type of imaging produces resolutions far better than methods that rely on focusing have been able to achieve. Other imaging methods, including several potential emerging methods such as multiple-input multiple-output (MIMO) antennas and confocal microwave imaging, also rely on various methods of processing the signals in ways that combine many signals, each taken at a slightly different time or location. By adjusting each one by an appropriate time or spatial delay, and combining many together, we can create a pattern that represents what would have happened if all of those signals had been physically focused on the point of interest. This creates a virtual image that reconstructs the desired image. Using signal processing to create a virtual focus is not unique to bioelectromagnetics. Radar, communications (such as your cell phone), and satellite imagery, to name a few, rely on some type of signal combining to produce the images or signals we desire. The future of bioelectromagnetic imaging and sensing is clearly headed in the direction that uses a combination of real and virtual localization to achieve better results.

Now let us look in more detail at two examples that demonstrate many of the principles in this book. Hyperthermia is a method of treating cancer by preferentially heating the tumor. Magnetic resonance imaging (MRI) is an imaging method that relies on strong magnetic fields to line up the spins in the hydrogen atoms in the body and coils to receive the very weak electric fields caused by their return to a normal state after being perturbed. Both of these bioelectromagnetic applications exhibit large differences in how they are applied, often as a result of the concepts described in the previous chapters.

6.3 Hyperthermia for Cancer Therapy

The topic of hyperthermia has been introduced earlier in this book as an example of the use of electromagnetic energy to heat tissues in the body. In this section, we discuss hyperthermia more fully and pay particular attention to those aspects that illustrate some electromagnetic concepts given in the first part of the book. We will not attempt to cover hyperthermia in great detail; there are numerous articles in the literature that discuss both the clinical and engineering aspects of hyperthermia. Here we use hyperthermia as an example of the advantages—and limitations—of employing electromagnetic energy for heating tissues. Application of heat for a wide variety of medical conditions is a very old medical therapy, but the use of electromagnetic fields to create this heat is relatively new. Cancer sites being considered for this therapy are breast, cervical, colorectal (spread to liver), endometrial, kidney, liver, lung, ovarian, pancreas, prostate, sarcomas (soft tissue cancers), and thyroid. Some European and Asian clinics are beginning to utilize hyperthermia in practice, but it is still largely experimental in the United States.

The use of heat to reduce or eliminate tumors has been known for over one hundred years, but only in the past few decades have the means for accomplishing reasonably controlled heating been available. Surface tumors (skin cancers, for instance) have been treated by burning or freezing them, and near-surface tumors have been treated by applying external heating (often hot baths). Use of artificially induced fever has also been examined for treatment of many diseases, albeit with limited or no success. Hyperthermia generally involves reaching and maintaining for several minutes a temperature of 42 to 45°C in the tumor. At the same time, the surrounding regions of the body must be kept cool enough that damage to them is minimal. In addition to killing the tumor by overheating, this method has been shown to increase the effectiveness of some types of chemotherapy drugs and radiation. In fact, heated chemotherapy drugs may be used to provide the heat directly to the location that the drugs are impacting. Sometimes this is done during surgery where heated chemotherapy drugs are circulated throughout the peritoneal cavity. This is called *continuous hyperthermic peritoneal perfusion* (CHPP). Clearly there are many ways of transmitting heat to the body, not all of which involve electromagnetics.

There are three main approaches to hyperthermia treatments. *Whole-body* hyperthermia raises the temperature of the entire body to nearly 42°C (about the highest that can be tolerated systemically). It is often uncomfortable for the patient due to temperature gradients, and tumors may not reach sufficiently high temperatures. *Regional* hyperthermia attempts to heat moderately large volumes, such as the thorax or pelvis, including the cancerous region as well as surrounding healthy tissue. The remainder of the body is kept as close to normal temperature as possible. *Localized* hyperthermia heats mainly the tumor itself (and perhaps the immediate border). It is used mainly for superficial tumors.

The first period of intense research in hyperthermia occurred during the 1970s and 1980s. Then, in the United States at least, there was a decrease in research activity, partly caused by the difficulty in obtaining reliable and uniform heating at deep tissue sites. In the late 1990s, there was a resurgence in hyperthermia interest, and new approaches and applicators were developed. Some of these new systems use ultrasound instead of electromagnetic energy for producing the heat, for reasons we will cover later in this section, and localized hyperthermia gained significant new attention.

The mechanisms of heat deposition in tissues by electromagnetic fields were described in Sections 1.6 and 1.16. When the tissue's electric dipoles (both permanent and induced) oscillate in response to the **E** field of an applied wave, heat is generated by a process analogous to friction. Similarly, when free charges (electrons and ions) in the tissue are set in motion by the **E** field, collisions with immobile atoms and molecules in the tissue generate heat. The propensity of the tissue to produce heat for a given sinusoidal **E** field magnitude is determined by the values of the imaginary part of its relative permittivity ε'' and its conductivity σ_c, or equivalently, by its effective conductivity σ_{eff}, as discussed in Section 1.14.

It is important to note that in each of the mechanisms above, it is the *internal **E** field* (i.e., the field inside the body) that is responsible for the heat generation. The external **E** field is important only to the extent that it causes the internal field. In addition, the internal **H** field is not directly responsible for heating, since tissue has a permeability μ close to that of free space, with no magnetic losses. Only when the **H** field produces a resulting internal **E** field—through a time-varying **H** field component, for example, as discussed in Section 1.4—will there be any tissue heating.

The frequency of the **E** field and the tissue properties determine the rate at which energy is given up to the tissue. The higher the frequency, the faster the wave will lose energy as it propagates through the tissue, and the shallower the heating will be. This is a key factor in determining the operating frequency for various applicators, as seen shortly.

Two of the most serious engineering challenges in hyperthermia are providing uniform heating throughout the target volume to ensure that all cancerous tissue reaches therapeutic temperature and is effectively treated, and achieving adequate temperatures in deep tumors without overheating the body surface or creating excessive temperature localizations or gradients in other regions of the body. Various designs of electromagnetic applicators have approached these challenges from different directions, and each has unique advantages, disadvantages, and regions of optimum application, which are generally governed by the laws of electromagnetics.

6.3.1 Types of Hyperthermia Applicators

Table 6.1 gives a listing of the various categories of applicators developed for electromagnetic hyperthermia. This list is by no means complete, and there are other possible ways of categorizing the applicators, but this one focuses on the mechanisms behind each type. The two major categories are: (1) *noninvasive* applicators, which use devices external to the

TABLE 6.1

General Types of Hyperthermia Applicators

Noninvasive (External Devices)	Invasive (Interstitial or Body Cavity)
Capacitive—low frequency	Electrodes
Inductive—low frequency	Radiative antennas
Radiative:	
Regional—low frequency	
Superficial—high frequency	

body to produce the internal **E** field, and (2) *invasive* applicators, which penetrate the body either through the skin or in natural body orifices. The noninvasive applicators are listed in order of the types of external fields that are principally responsible for the internal **E** field.

6.3.1.1 Capacitive Applicators

In this type of noninvasive applicator, two conducting electrodes are placed on or near the surface of the body. The electrodes can have various shapes and sizes. One simple example is shown in Figure 6.1. A voltage source is connected across the electrodes, producing an **E** field stretching throughout the volume between them. Since these applicators are often intended to heat deeper tissues, the frequency of the voltage source is relatively low (in the high kilohertz to low megahertz range). Therefore, as explained in Section 2.2, **E** and **H** are uncoupled and essentially no **H** field is produced; this is a predominantly **E** field applicator. The **E** field lines terminate on charges contained on the surface of the electrodes.

The advantages of the capacitive-type applicator are based upon its simplicity. The placement and shape of the electrodes can be tailored to the location of the region that is to be heated. It is relatively easy to visualize the paths that the field lines take. Also, the electrodes can be curved to match the skin's contour.

Unfortunately, however, the fields generated in the tissue are not optimum for preferentially heating deep tumors. As can be seen in Figure 6.1, or in any other arrangement that can be envisioned using capacitive-type electrodes, the **E** fields are mostly perpendicular

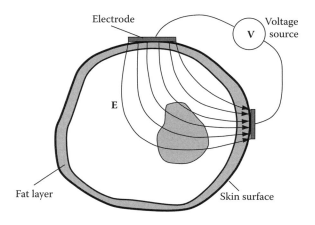

FIGURE 6.1
Simple example of a capacitive applicator in which two electrodes are placed on the surface of the skin. The **E** field lines shown are not exact, but are representative of those expected from a capacitive-type applicator. They are mostly perpendicular to the skin's surface.

to the body's surface, where there are fat layers. If muscle or muscle-like tissue is beneath the fat, the fields are mostly perpendicular to the fat-muscle interface. As explained in Section 2.4, the boundary conditions for normal **E** field components combined with the lower permittivity of fat means that the **E** field in the fat is much higher than in the muscle. Even though fat is less lossy, the higher **E** field results in higher energy deposition, and often overheating of the fat layer. A common tendency with capacitive applicators is to burn areas on the surface of the body when attempting to heat deeper tissue.

This problem is compounded by the inclination of **E** fields to concentrate at the edges of metallic electrodes; this effect was shown in Figure 2.8. Again, skin near the corners of the applicator is vulnerable to burns. To reduce this problem by spreading out the fields, water boluses (nonmetallic containers of water) are sometimes placed between the electrodes and the skin. The water in the bolus can even be chilled and recirculated to provide some conductive cooling of the skin.

6.3.1.2 Inductive Applicators

In this type of applicator, an external coil (or some other means of generating high currents near the body) is used to produce an **H** field inside the body. As mentioned above, the magnetic field itself will do no heating, but if the **H** field varies with time, it will induce an internal **E** field for heating. Since external **E** fields are usually not desired with these applicators (for the reasons stated above and also to make coupling to the body less sensitive to positioning), they are usually operated at low to moderate frequencies. This keeps the external **H** field from generating a significant external **E** field, as discussed in Section 2.2. In addition, these applicators are generally meant for deep heating, again suggesting a lower frequency. However, since the generation of the internal **E** field is proportional to the time rate of change of the **H** field (see Equation 1.3), the frequency should be high enough to produce a sufficient internal **E** field. Operating frequencies are generally in the low megahertz range, such as 13 or 27 MHz.

Figure 6.2 shows the simple case of a coil surrounding a cylindrical body. The **H** field lines run longitudinally through the body, then encircle the coil outside the body since **H** field lines must close upon themselves. Depending upon the exact geometry of the coil windings

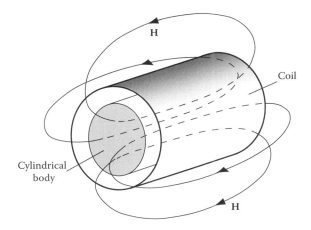

FIGURE 6.2
Simple example of an inductive applicator for hyperthermia. In this example, the coil extends around the entire body, but in other forms, the coil can be saddle shaped or pancake shaped.

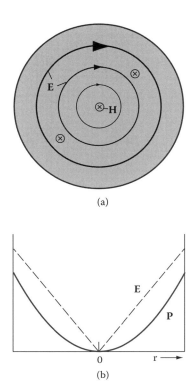

(a)

(b)

FIGURE 6.3

(a) The **E** field lines inside a homogeneous cylindrical body placed in the coil of Figure 6.2. Note that the **E** field lines (and thus the eddy currents) circulate around the center, where they are zero, and grow radially toward the periphery. (b) A plot of both the **E** field magnitude and power deposition pattern P. This shows the unequal heating expected from this type of applicator.

and the size of the body, the density of the **H** fields inside any cross section of the body can be fairly uniform. Other forms of coil applicators can also be used, such as pancake coils or saddle-shaped coils, or specially shaped conductors that bring currents close to the surface of the body. All inductive applicators share common advantages and disadvantages.

One advantage of this type of applicator is its relative insensitivity to the coupling conditions. Since tissue is nonmagnetic, the exact position of body with respect to the coil does not affect the **H**-field pattern. Since the external **E** fields are weak, the permittivity of the tissues also has little effect on the coupling. Therefore, as the patient moves, the coupling of energy is relatively unaffected. In addition, the tuning of the coil in the resonant electrical circuit of the source is forgiving of exact body positioning.

One disadvantage shared by almost all inductive-type applicators is that the pattern of the induced internal **E** field is not optimum for deep central heating. This is shown in Figure 6.3 for the simple body model of Figure 6.2 consisting of a cylinder of uniform tissue properties. Since the **E** field lines are produced by the time-varying **H** field, they encircle the **H** field lines. There is a center of rotation for the **E** field (in this case it is common to the center of the cylinder), and at this point the field is zero. The tissue conduction current caused by the **E** field (called the eddy current) is also zero here. The field and current grow linearly toward the periphery of the cylinder. Because power deposition is proportional to the square of the **E** field (see Equation 1.41), the heating pattern has a parabolic shape.

The situation illustrated in Figures 2.40 and 2.41 for a culture dish exposed to a transverse magnetic field displayed identical behavior.

Figure 6.3(b) plots both the **E** field pattern and the power deposition pattern. Clearly, centrally located tumors would not be heated effectively. Heating is greatest at the periphery, showing that surface heating is a major concern with inductive applicators, as it is with capacitive applicators.

If the tissue properties are not uniform as in this simple example, the eddy currents will not follow a radially linear profile and will be more irregular. This sometimes can be used to advantage. For example, a high-conductivity tumor surrounded by lower-conductivity tissue will have a local eddy-current pattern flowing around the approximate center of the tumor. This is similar to the configurations shown in Figures 2.42 to 2.44 for a higher-conductivity object placed in a saline culture dish. The local eddy-current patterns can lead to increased heating of a deep tumor, but the amount of improvement depends upon the conductivities of the tissues involved, which may vary considerably from case to case.

When other coil shapes are used, their **H** field patterns will be different from the simple case in Figure 6.2. The orientation and extent of the **H** fields can be manipulated by the size and position of the coil. However, the basic constraints to the eddy current and heating patterns given above still hold. The current must still circle about some center where it is zero, and therefore there will be a null in the heating pattern. The current will be larger toward the periphery, and surface overheating must be watched for.

6.3.1.3 Radiative Applicators

This class of applicators relies upon the coupling of **E** and **H** together to carry electromagnetic energy into the tissue. They operate either at higher frequencies, when localized surface heating is needed, or at lower frequencies, when deeper penetration is desired. The applicator and feed configurations are chosen to maximize the coupling of the launched wave into the tissues.

One version of a radiative applicator is shown in Figure 6.4. It basically consists of an open-ended waveguide that is coupled to the skin with a quarter-wavelength matching slab. (See Section 3.3.2 for a discussion of how reflections are affected by the permittivity

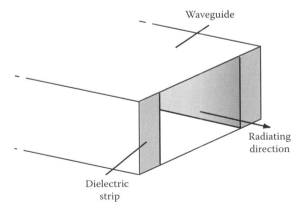

FIGURE 6.4
A radiative hyperthermia applicator consisting of an open-ended waveguide. The waveguide is loaded with two dielectric strips on either side to cause the **E** field mode pattern to be more uniform in the transverse direction.

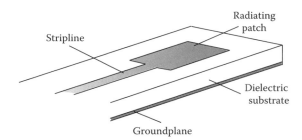

FIGURE 6.5
A radiating patch applicator. The applicator is lightweight and small, making it more convenient. It is intended for heating superficial tumors.

and thickness of a slab. The permittivity of the quarter-wavelength matching slab is chosen to be equal to the geometric mean of the permittivities of the media on either side of the slab, which minimizes reflections.) The waveguide is loaded on both sides with dielectric strips; this produces a mode structure that approximates a TEM mode (as described in Section 3.5.1), thus giving a more uniform pattern in the transverse direction than an unloaded waveguide. The size of the waveguide dictates its relatively high operating frequency, 2,450 MHz, so it is appropriate for heating superficial tumors.

To make the applicator size more compact, microstrip radiators have also been developed. One style, a patch radiator at the end of a microstrip transmission line, is shown in Figure 6.5. These applicators are lightweight and can even be made flexible, so they are more convenient to use than the larger, heavier waveguides. They operate at higher frequencies (433 to 2,450 MHz), so are meant for localized superficial heating.

Larger, lower-frequency radiators have also been designed for heating larger tissue volumes. Some surround the body with an array of several radiating apertures. The intent here is to penetrate the body to deeper locations for regional heating, so they use frequencies that are lower than the smaller devices (around 100 MHz).

Deep body hyperthermia can be accomplished with an array of antennas outside the body that focus energy on the tumor location inside the body. If it is desired to heat the entire torso, this could be done by sequentially focusing on subregions of the torso. An annular phased array is shown in Figure 6.6. The antennas (which are oriented longitudinally along the body) are made of conducting plastic so that they can be used in an MRI machine. The whole device is filled with water and includes probes to measure temperature inside the body while the hyperthermia treatment is being performed.

All electromagnetic radiative applicators face the same trade-offs among depth of penetration, applicator size, and localizing ability. The foundations for these trade-offs have already been introduced in earlier sections. Figure 6.7 shows the penetration characteristics for planewaves of various frequencies into a dielectric halfspace whose properties are similar to those of muscle. The higher-frequency waves are clearly attenuated quickly by the tissue due to their high loss. Although the waves coming from practical applicators are not planewaves, and the body certainly is not an infinite halfspace, this same general behavior is expected to apply to radiative applicators. Note from Figure 6.7 (and Figure 1.40) that in order to penetrate to reasonable depths (say beyond 7 or 8 cm), the frequency must be about 100 MHz or lower. But at 100 MHz, the relative permittivity ε'_r of muscle is about 100 and the conductivity σ is approximately 1 S/m (see Figure A.1). The wavelength in muscle, therefore, is quite large, about 30 cm.

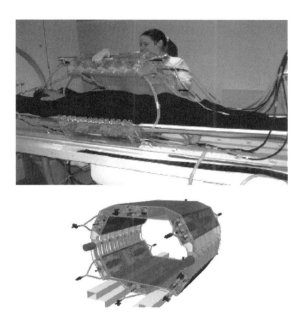

FIGURE 6.6
A prototype of the new Berlin MR-compatible water-coated antenna applicator (WACOA). (From Nadobny, J., et al., *IEEE Trans. Biomed. Eng.*, 52/53, 505–19, 2005. © IEEE. With permission.)

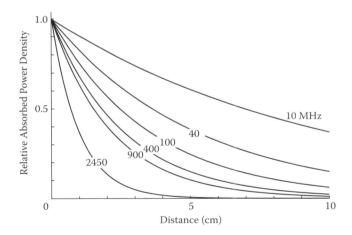

FIGURE 6.7
The penetration of planewaves of various frequencies into a dielectric halfspace with the properties of muscle. There is significant attenuation of the higher frequencies. To achieve deep penetration (beyond 7 or 8 cm), frequencies of 100 MHz or lower must be used.

Now, Section 3.7 explained that a radiator is not very effective unless at least one dimension of the radiating structure is one-half of a wavelength or larger. The wavelength referred to is the wavelength in the medium into which the wave is radiated, in this case tissue. Therefore, to be efficient, the 100 MHz radiator must be 30/2 = 15 cm in size or larger. If the frequency is lower, the applicator will be even larger. This means it will be rather bulky and heavy, and more importantly, the energy coming from the applicator will spread out due to diffraction. Localized heating is difficult at the low frequencies that will penetrate deeply.

In an attempt to get deep localized heating, the applicator can be made smaller and perhaps focused, while keeping the same low frequency. But again, the laws of electromagnetic propagation work against this. Not only will the smaller radiator be less efficient, but the beam will diverge more rapidly. This was covered in Section 3.8, which showed that the diffraction of a wave away from a finite aperture is inversely proportional to the ratio of aperture size to wavelength. When the aperture is a fraction of a wavelength in length, the divergence is very large, and no deep localization is possible.

Even if a focusing lens is used, Equation 4.21 in Section 4.4.2 illustrates that focusing is futile here. (Equation 4.21 is actually valid only for gaussian beams, but can be used here as an approximation.) For practical reasons—the extreme curvature required and the resulting aberrations—it is very difficult for any lens (microwave or optical) to focus an electromagnetic wave to a point closer to the lens than the diameter of the initial beam. Thus, in Equation 4.21, f/D is rarely smaller than unity, and the size of the focused beam, approximately given by d_0, is on the order of a wavelength or larger. Since the wavelength is already large for low-frequency waves, the heated region is large, and again, no deep localization is possible.

Ultrasound waves, described in Section 3.9, obey these same laws, but with different constants and with a much different outcome. The speed of sound in tissue (1.5×10^5 cm/s) is much smaller than the speed of electromagnetic radiation (approximately 3×10^9 cm/s in average tissue at 100 MHz). The frequencies that are appropriate for depositing power at deep levels (10 cm and deeper) are also lower with ultrasound, between 100 kHz and 1 MHz, due to ultrasound's different absorption coefficients. For an ultrasound beam at 500 kHz, for example, the wavelength in tissue is only $(1.5 \times 10^5)/(500 \times 10^3) = 0.3$ cm. Therefore, an ultrasound beam can simultaneously deposit heat at deep tissue locations *and* be focused to a small spot. In fact, the focused ultrasound beam is usually too small to give uniform coverage of the treated area, so it is either purposely defocused or swept around to get the desired volume coverage.

Ultrasound's advantage of being able to penetrate deeply with small-wavelength beams is one reason ultrasound energy is being seriously considered for hyperthermia therapy. A disadvantage, however, is that ultrasound will not effectively penetrate bone or air, so treatment is limited to regions of the body where access is through soft tissue. Also, the temperature gradients potentially present at bone interfaces, in particular, are problematic.

6.3.1.4 Invasive Applicators

In order to circumvent the difficulty of obtaining deep, localized heating patterns from external electromagnetic applicators, invasive probes may be used. These probes are placed in natural cavities of the body (if the tumor is nearby) or directly through the skin. Needles used to inject chemotherapy drugs can double as invasive applicators for the tumor. Postsurgical drains can also act as conduits for invasive therapy.

Interstitial hyperthermia typically requires placing one or more monopole antennas within the tumor or in the region around the tumor. Coaxial cables with the center conductor extending beyond the outer ground shield of the cable make simple interstitial antennas, as shown in Figure 6.8, and have been used for treatment of highly localized tumors, such as in the prostate. One problem with these antennas is that they have a teardrop-

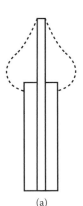

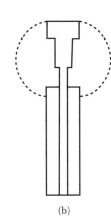

FIGURE 6.8
Interstitial applicator monopole antennas made from coaxial cables where (a) the inner conductor is a uniform thickness and (b) the inner conductor increases in thickness. The radiation pattern (which can be thought of as heating pattern) is a teardrop shape for (a) and is a more uniform ellipse for (b). No power is transmitted out of the tip of the antenna, as described in Section 3.7.

(a)   (b)

shaped radiation pattern (Figure 6.8(a)), which means there is little or no heating near the tip of the antenna. The needle-shaped applicator would then have to be extended through and past the tumor in order to heat it. To make the heating pattern more uniform, the inner conductor that is protruding from the coaxial line can be made thicker at the end, as shown in Figure 6.8(b). Interstitial applicators are also used in body orifices. When many interstitial antennas are used, often circling the tumor, their feed systems can be phased (with different delays between them) in order to use the constructive interference produced to steer the maximum heating location. This is still not commonly done in practice, because of the difficulty (discomfort, infection, and fear of reducing the ability of the body to naturally contain the tumor) of placing multiple needle applicators around the tumor. Phased-array antennas usually require higher frequencies (for example, 915 or 2,450 MHz). Alternatively, lower frequencies can be used with needle applicators. These then become simple electrodes and provide heating by supplying conduction current. A coaxial line can also be cut off (with the center conductor not extending past the ground shield) and pressed against the tumor. The field lines are much the same as those described for measuring the electrical properties of tissue in Section 3.3.2 and result in heating at the tip of the coax.

The advantage of invasive probes is that the heat can be localized with more precision and in a smaller volume at depth than with external applicators. One disadvantage, of course, is that it is much more uncomfortable for the patient. Also, even using multiple probes does not ensure uniform heating; there still may be considerable nonuniformity to the power deposition pattern, depending on the placement and individual patterns from the probes.

6.3.2 Engineering Problems Remaining in Hyperthermia

It should be clear from the previous discussion that one area that remains problematic with electromagnetic hyperthermia is the ability to heat deeply in a well-controlled and localized manner. Too often there is surface overheating that accompanies deep heating, regardless of the type of applicator used. Based upon the concepts of electromagnetics, it seems unlikely that this problem will be easily solved. When localized superficial heating is desired, on the other hand, several of the approaches, in particular the small radiating applicators, are successful and will probably continue to expand in usability.

Other engineering issues remain. These include the need for multiple-point temperature measurements for accurate and thorough monitoring. Treatment planning will require accurate characterization of the applicator deposition pattern and the tissue parameters, as well as a numerical technique to predict the resultant heating pattern. Tissue perfusion significantly modifies the temperature distribution for any given power deposition pattern, often in a time-variable and unpredictable way. Still, the promise of even a partially successful therapy for cancer spurs the continued study of hyperthermia.

6.4 Magnetic Effects

6.4.1 Magnetic Resonance Imaging (MRI)

Hyperthermia is one example where we can see that focusing electromagnetic fields at specific points inside the body using sources that are on the outside of the body is very difficult and often practically impossible.

Nevertheless, MRI scanners can create submillimeter-resolution images inside the body using magnetic fields produced by an external source. Magnetic resonance imaging (MRI) is a painless, noninvasive way of imaging soft tissues inside the body. MRI uses a very strong DC magnetic field (0.3T, 0.5T, 1.0T, 1.5T, 3T, 4T, 7T or even greater in some applications) to make the magnetic dipoles of the nuclei in the hydrogen atoms in the body line up such that their spin directions are aligned parallel to this longitudinal main magnetic field. This is similar to what a compass needle would do in a strong magnetic field. Separate coils (called switched gradient coils) superimpose a weaker, but spatially varying, magnetic field on the main field, resulting in an overall magnetic field with precisely controlled spatial variation in strength in multiple directions. Then an RF transmit coil imposes a short RF transverse magnetic field on these dipoles to cause them to tilt away from their original orientation (the maximum signal is obtained at 90°, but smaller angles are used for faster images). At the conclusion of the RF pulse, the dipoles relax back to their original equilibrium position in the longitudinal direction at a specific rate or frequency, known as the Larmor frequency, proportional to the DC magnetic field strength at the particular location of the hydrogen atom. The Larmor frequency for hydrogen nuclei is 42.58 MHz per each Tesla of field strength; thus for a nominal 3T magnetic field, the Larmor frequency is 127.74 MHz (but will vary proportionally to the exact field strength as determined by the gradient coils). A set of receiver coils is able to detect the RF electromagnetic field created when these magnetic dipoles return to their normal orientation with frequencies (and phases) unique to the location of the atoms in the body. The magnetic relaxation properties of the different tissues affect the relative duration of the local received signals and with a carefully designed pattern of switched gradients, a 3D map of the body's tissues can be produced.

Modern clinical scanners typically use a superconducting electromagnet to create the very strong magnetic field along the major axis of the body (head to toe). During an MRI scan, the patient is placed on a nonmagnetic table that is moved in or out of the bore of the magnet. Because the body is essentially nonmagnetic (the permeability is essentially that of free space; see Section 1.6), it does not affect the design of the gradient coils, which greatly simplifies the design. Furthermore, for the coils that produce the pulsed RF transverse magnetic field of the scanner, the E field in the body is only very weakly coupled to the B field at the frequencies normally associated with lower strength magnets. Therefore the magnetic-field coils can be designed on the basis of producing the desired RF magnetic field in free space.

Typical receiver coil designs are focused on one of two competing criteria. One option is to make the coil have a homogeneous imaging volume. These coils are usually larger volume coils that surround the patient's anatomy of interest. The coils help provide consistent tissue image intensities throughout the volume of interest. The second type of coil design is optimized for increased signal-to-noise ratio (SNR). These coils are constructed for localized imaging of a specific region of anatomy, and usually comprise one or more single-loop surface coils. It is important to realize that homogeneity and SNR cannot both be simultaneously optimized for a given receiver coil.

Since MRI spatial resolution is determined by a spatially varying magnetic field that encodes location as a unique frequency and phase of the received signal, the difficulties of localizing internal EM fields described in the previous section are not encountered in MRI scanners. The problem, instead, is to produce carefully controlled gradient B fields that are superimposed on the static magnetic field. Although coils have been designed for producing satisfactorily homogeneous RF transverse magnetic fields in the 40 to 60 MHz frequency range of typical MRI systems, difficulties have been encountered as the frequency of advanced systems has been increased with stronger DC fields to achieve increased resolution. As the frequency is increased (for magnetic fields above 3T for example), the coupling of the B field to the E field in the body increases, and also the wavelength inside the body becomes smaller. Both effects cause increased interaction between the B field and the body, making it difficult to achieve homogeneous RF magnetic field excitation; this effect is sometimes referred to as a standing wave phenomenon. Special techniques are required to overcome these difficulties at higher frequencies.

Like hyperthermia applicators, the RF receiving coils can be designed for the whole body, a region of the body such as the head or a limb, or a localized region such as the neck only. The objective of the receiving coil is to pick up as much signal from the region of interest as possible while minimizing the noise. This signal-to-noise ratio (SNR) determines the quality of the final image. The amount of signal a coil receives is proportional to the volume of interest (the voxel size) in the tissue region and the sensitivity of the coil in this voxel (a voxel is the smallest volume segment imaged by the MRI scanner). The closer the coil is to the voxel of interest the stronger the signal induced in the coil according to Faraday's law. If we are interested in imaging the blood vessels of the neck, then these constitute the volume that determines the strength of the signal. However, the noise is roughly proportional to the overall size of the coil imaging volume. So the noise would be largest for a whole body coil, smaller for a regional coil, and smallest for a localized coil. Thus, using a coil that is as small as possible while being large enough to see the overall region of interest gives the best possible signal-to-noise ratio. This concept is the basis of many improvements in MRI images.

There are three basic types of receiver coils used for MRI. Volume coils, such as the quadrature birdcage head coil shown in Figure 6.9, have been used for imaging large and deep anatomic structures of the body and for providing homogeneous field profiles (but they are being supplanted by phased arrays—see below).

For high-resolution applications that are more localized, such as hippocampus imaging, cochlear imaging, and functional imaging in which the object features are very small, volume coils pick up less signal and more noise because they are far removed from the region of interest. For imaging small, detailed regions of the body, smaller surface coils

FIGURE 6.9
Quadrature birdcage coil with endcap used for whole-volume head imaging. (From Hadley, J. R., et al., *J. Magn. Reson. Imaging*, 11, 458–68, © 2000. With permission.)

have improved image quality, since they pick up only the signals of interest and (because they are smaller) pick up less noise.

Modern technology now allows several individual small coils to be combined into phased arrays (PA) such as the one shown in Figure 6.10, where each coil picks up the signal slightly out of phase from the other coils. The data are then combined, giving the best possible signal-to-noise ratio for the application. Part of the price for the improved image quality seen with phased-array coils is the complexity of the receiver and data acquisition system, since each antenna is received on an individual channel. A particular MRI scanner has a given number of channels for reception, limiting the number of coils in the phased array. The image processing is also more computationally expensive, as the signal from each antenna is weighted depending on its proximity to the target region (and hence the

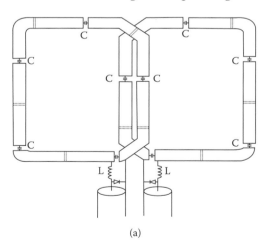

(a)

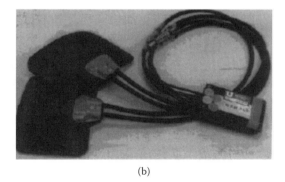

(b)

FIGURE 6.10
(a) Two-element phased-array coil design. Dashed lines indicate the breaks in the underside of the double-sided copper section of the coil. (b) Image of finished phased-array coils (enclosed in foam) with triax balun cables and phased-array port connector box. (From Hadley, J. R., et al., *J. Magn. Reson. Imaging*, 11, 458–68, © 2000. With permission.)

expected relative SNR), phase shifted, and combined with the other similarly processed signals.

The field of view of the PA depends on the number of channels available. With an increased number of channels, a larger region of interest can be covered. Also, overall image acquisition time for a PA can be reduced because of the increased SNR available from the PA coils; SNR can be traded for imaging speed, resolution, or both.

MRI is still a rapidly advancing field. Use of specialized partial-body coils, higher Tesla magnetic fields, and stronger, more localized field gradients is providing better resolution and detail in MR images. For instance, imaging of blood vessels and arteries is reaching levels of precision that can be used for prognostic intervention as well as guidance of vascular surgery. MRI is used during many surgical procedures, including those in the brain and vascular system, removal of tumors, and laparoscopic procedures. Functional MRI, which uses the contrast between oxygenated hemoglobin and deoxygenated hemoglobin to show neural activity in the brain, for instance, relies on some of the most sensitive MR imaging available. Here specialized head coils are used to improve the signal-to-noise ratio. MRI can also be used to precisely image temperature within the body and is therefore useful for guiding ultrasound or RF hyperthermia treatments. It is highly likely that future advances in MR imaging will produce faster, higher-resolution images of the body, particularly for specialized applications and specific areas of the body. Use of MRI in surgery will almost surely continue to grow, and functional MR imaging is likely to impact our ability to diagnose and treat diseases.

6.4.2 Nuclear Magnetic Resonance (NMR) Spectroscopy

Magnetic resonance (MR) can also be used for spectroscopy where individual chemical components have different MR resonant frequencies that will give a map of the chemical composition. In the future it is possible that this type of spectroscopy may also be applied within the body, though challenging, to determine the composition of various gases or chemicals within the body for the diagnosis of biological conditions. This is called multinuclear imaging.

Nuclear magnetic resonance (NMR) spectroscopy has been used extensively to determine the chemical composition of many fluids, including biological fluids (such as blood and urine). A strong static magnetic field aligns the spins in the various nuclei in the fluid. In one common configuration, a pulsed RF electromagnetic field transversely varies this magnetic field, causing the chemical species in the fluid to resonate at their natural Larmor frequencies. The Larmor frequency depends on the chemical signature of the fluid and the strength of the static magnetic field. The resonating nuclei induce a signal called the *free induction decay* (FID) in a receiving RF coil. The Fourier transform of the free induction decay gives a spectrum of the chemical resonances. Each individual chemical has a unique resonance or series of resonances, which can then be used to determine its existence and abundance in the mixture.

The sensitivity of an NMR experiment is determined by the strength of the static magnetic field. It is critical that this field be as uniform as possible across the sample of interest, since the frequency of the resonances is proportional to the field strength. Therefore, one desires as strong and uniform a field as possible. Spectrometers today use electromagnets, as described in Section 1.5. However, there exists the possibility of using rare-earth permanent magnets as a good alternative to the power-intensive electromagnets in terms of portability and cost of operation. Magnets with strengths on the order of 1-3 Tesla have been demonstrated and could provide sufficient static field for effective NMR. A permanent

FIGURE 6.11
Please see color insert following page 146. A millimeter-resolution NMR coil wrapped around a needle. The needle and coil are then encapsulated in plastic or rubber, and the needle is removed. This creates a 100% fill-factor coil. (From J. Stephenson.)

magnet NMR system might open the door for application in developing countries and academics, where cost has been prohibitive in the past. Applications such as bio-warfare agent detection and point-of-care or doctor's office testing are foreseeable.

In addition to the strong, uniform static magnetic field, the design of the receiving coil is very important. The signal-to-noise ratio determines the effectiveness of NMR spectroscopy. The signal is proportional to the amount of sample inside the receiving coil. The noise is determined by the volume of the coil. This means that anything that is not sample inside the coil contributes only to the noise and not to the signal. Microscale NMR coils are typically fabricated using quartz capillary tubes, which are heated and pulled to a desired diameter. Copper wire is then wound around the capillary tube by hand or machine. The disadvantage of this method is that the quartz tube fills a portion of the NMR coil, and as a result reduces the volume of the coil interior that the sample can fill. In microscale NMR experiments the signal is very small to begin with because of the tiny amounts of sample being analyzed. An alternative method of producing NMR coils uses a needle wrapped with a wire, encapsulated in near-solid plastic, after which the needle is removed. This leaves a channel for the fluid that completely fills the coil and increases the signal-to-noise ratio. One such coil is shown in Figure 6.11. This and improved receiving hardware may soon enable microscale NMR in the small, low-cost package that permanent magnets could provide.

6.5 Proposed Bioelectromagnetic Effects

Section 1.18 described a number of well-recognized biological effects caused by electromagnetic waves. There are several other effects that have been proposed but are still in the research and exploratory stage. This section is devoted to those effects that are plausible but have not yet been either proven or disproven as potential sources of bioelectromagnetic effects.

6.5.1 Soliton Mechanisms

It is well known that direct stimulation of the cell can trigger chemical events within the cell. The soliton theory proposes that external electric fields can create a nonlinear soliton wave in the proteins associated with the cell membrane channels. This wave could then propagate along the protein and through the membrane, thus remotely triggering enough energy to create a chemical response within the cell and resultant biological effects.

6.5.2 Spatial/Temporal Cellular Integration

Some experiments claim to show biological response when the electromagnetic fields are below the naturally occurring fields. A proposed explanation for these observations is that the very small externally applied fields may be spatially or temporally correlated, whereas the endogenous fields are not. Thus, the endogenous fields would not add when integrated over time, whereas the externally applied fields might. This effect is difficult to measure due to the very small field strengths being considered.

6.5.3 Stochastic Resonance

This theory follows the well-known electronic effect where noise in a bistable or multistable system can augment a weak signal. It is theorized that the cellular membrane may be sufficiently nonlinear to allow this effect to occur within the cell.

6.5.4 Temperature-Mediated Alteration of Membrane Ionic Transport

This theory is based on the observation that one biological mechanism for producing changes in cardiac response is a very small, brief transient temperature spike. Having observed this mechanism for higher field strengths, it was theorized that much smaller field strengths might be capable of producing similar temperature spikes on cellular membranes.

6.5.5 Plasmon Resonance Mechanisms

Plasmons are sheets of surface charge, in this case on the cellular membrane. The charges can build up at the junctions between the insulating membrane and the conductive inner portion of the cell. It has been theorized that external electromagnetic fields in the 1–10 GHz range could cause these plasmons to resonate, thus affecting membrane ion transport and conformation of the intramembrane proteins and lipid bilayers. Changes in these cell parameters could lead to pathogenic consequences.

6.5.6 Radon Decay Product Attractors

When radon, a radioactive gas, decays, it produces aerosols that are attracted to common sources of electric power such as appliances and power lines. It has been theorized that these decay products might also congregate in the human respiratory tract when a person is exposed to even relatively weak (1 kV/m) electric fields, thus increasing his or her risk of cancer.

6.5.7 Rectification by Cellular Membranes

Nonlinear systems such as proteins in cellular membranes have been observed to rectify strong oscillating alternating current (AC) fields so that a small direct current (DC) potential bias can develop from them, thus adjusting the cellular membrane potential. The percentage of rectification is very small, and it is not clear if this effect is biologically significant for weaker electric fields.

6.5.8 Ion Resonance

This theory suggests that an ion placed in a static magnetic field (such as the earth's magnetic field) will circle around the field's vector. If an alternating electric field is also present, the ion will resonate. This resonance can change the energy states of the ion and affect cell biological activity in cultures. It is not yet clear if this effect can be extended to probable excitation sources or animal systems.

6.5.9 Ca^{++} Oscillations

Ca^{++} ions send information to the cell coded in their resonant frequencies. These resonances may be altered by extremely low frequency (ELF) electric fields surrounding the cells, and could therefore potentially alter the biological effects within the cell.

6.5.10 Magnetite Interactions

Ferromagnetic compounds known as magnetite are found in the brains of migratory birds. These compounds increase the birds' sensitivity to the earth's magnetic field and facilitate navigation. Traces of magnetite are also found in the human brain and possibly in other tissues as well. It is proposed that weak magnetic fields could interact with the magnetite either through DC forces or through microwave resonances that might enhance energy absorption in the tissues.

6.6 Emerging Bioelectromagnetic Applications

The advantage of electromagnetic fields is that they can be transmitted, guided, and focused. This can be done noninvasively, semi-invasively, or highly invasively. They can transmit power (to be used for heating) or data (to be used for communication). They can also receive signals that are produced naturally by the body or that are induced by a transmitted power and then altered by the body. This ability to interact with their remote environment is probably the greatest strength and also the greatest challenge for electromagnetic fields. The present and future potential of bioelectromagnetic applications depends on how well we are able to control the propagation of the waves into and out of the body. This final section shows how our ability to become more and more localized in this control is opening up the newest bioelectromagnetic frontiers.

6.6.1 Low-Frequency Applications

Chapter 2 described the wide range of applications for low-frequency electromagnetic fields where the wavelength is very large compared to the body. In this realm, the fields are most often induced or received by electrodes and are treated as voltages and currents on these electrodes. Many applications are evident today, including cardiac pacemakers and defibrillators, pulsed electromagnetic nerve stimulation for treatment of bone and soft tissue injuries and for pain control, and direct nerve stimulation for treatment of Parkinson's disease and others, and diagnosis of neurological dysfunction and injury. The most advanced applications today rely on small electrodes implanted in the body, and the future of low-frequency bioelectromagnetics is moving toward even smaller electrodes. Sub-millimeter-sized electrodes are now available that can stimulate or receive impulses from a single neuron. These tiny electrodes can be used individually, but combining them into arrays of electrodes, each precisely stimulating an individual neuron, holds even more promise. Retinal and cochlear implants are under development and have already shown that this technology is basically feasible.

One of the most interesting challenges that remain in this development is understanding how the visual and auditory systems truly function at a microscopic level, and how electrical signals on the nerves stimulate the vision and sound we experience. These highly nonlinear and nonuniform systems provide an amazing window into how the brain functions, and the tiny electrodes that have become available in just the last decade hold the key to some of the most intriguing brain research opportunities available today. Other practical challenges in the development of neural prosthetics include miniaturized packaging (including what to do with the unavoidable heat generated by the electronics), long-term biocompatible materials at this scale, efficient data communication and power transfer (battery recharging), and surgical implantation methods that minimize damage to the nerves of interest.

In the future, it is possible that many neurological disorders and injuries may be treated by miniaturized electrodes implanted in the body. For spinal cord injury, these electrodes may pick up signals on one side of the break and transmit them to electrodes on the other side, thus completing the broken circuit. It may even be possible to stimulate and control the healing and regeneration of nerve tissue using electrical signals. The electrical frontier for understanding brain and neurological function is rapidly advancing because of the ability to receive and transmit more and more localized electrical signals.

As this technology progresses, it promises to bring with it a far more advanced understanding of how we think, feel, see, hear, taste, and move. This understanding will form the foundation for the coming generations of bioelectromagnetic devices. Many of the treatment devices used today apply pulsed electromagnetic fields with one of many different waveforms. These waveforms are generally developed experimentally using a try-and-see approach. A more advanced understanding of exactly how our bodies work is likely to provide a tremendous opportunity for advancements in signal processing for improved treatment and diagnosis.

6.6.2 Medium-Frequency Applications

Chapter 3 described a wide range of applications for medium-frequency electromagnetic fields when the wavelength is on the order of the size of the body. In this region, the fields can be focused using both internal and external applicators with reasonable success. Hyperthermia treatment for cancer and other heating applications, communication systems for implantable devices, and MRI are among the most successful applications in this region. Their success today and progress in the future rely on more localized application and reception of the electromagnetic fields. Earlier research concentrated on physical ways to accomplish this focusing, and in fact, much work continues in this area today. Localized MRI coils for imaging the arteries in the neck and the breast, for instance, have significantly enhanced images of these regions. Localized coils for functional imaging of the brain are also interesting applications of this method. Going to higher and higher frequencies helps to localize the fields; however, it also reduces the ability of the fields to penetrate the body. Modern MRI machines use increasingly higher magnetic fields, which prompts questions and challenges, including how to implement more localized coil designs, how to minimize peripheral nerve stimulation, and how to image regions that include metallic implants such as pacemakers and the pins and plates often used to repair broken bones.

Signal processing and image processing are two critical research areas that are helping to overcome the physical limitations and propel bioelectromagnetic imaging into the next generation. Combining signals obtained from many receivers at different frequencies and times has made subwavelength imaging a reality, and great strides continue to be made. This same signal processing is being utilized to better understand the endogenous signals received from the body. Ultrasound imaging is one example where this has been applied to see real-time body function from external measurements using advanced signal processing. Similar methods are being applied to microwave imaging and MRI. Electroencephalography (EEG), for instance, receives signals from the brain and processes them to better understand brain function. Enhanced signal processing is drawing more and more information from these signals. Conformal microwave imaging and microwave tomography for breast cancer imaging both rely on signal processing to combine the received signals in ways that enhance the ability to interpret and understand them. MRI and CT imaging continue to improve due to advanced signal and image processing algorithms.

Dosimetry, described in Chapter 5, is another key factor in mid-frequency bioelectromagnetic research. The ability to predict how the fields interact with the complex heterogeneous body has improved greatly over the past decade and is being applied in the design of the electromagnetic devices, in prediction of how the endogenous fields behave in the body, and in algorithms used for tomographic imaging. The dosimetric methods for analyzing the body at millimeter resolution are now well established, although improvements in speed and efficiency continue to expand the applications where they can be applied in practice (particularly for imaging). Similar methods are used for modeling individual cells and neural systems within the body, and a significant area of bioelectromagnetics research is how to model more localized phenomena. Another area of dosimetry that is likely to add to the future understanding of how our bodies function is the modeling of multiple aspects of the system simultaneously. For instance, modeling electromagnetic fields and heating together has been very important to better understand hyperthermia and heat dissipation from implantable electronics. The ability to model electromagnetic fields, heating, chemical changes, and biological responses to these and other impact factors may help explain some of the little-understood effects seen today.

6.6.3 High-Frequency Applications

High-frequency electromagnetic applications using infrared, visible light, or x-rays were described in Chapter 4. In this realm, the fields behave as rays and can be modeled as such. The use of the natural infrared radiation from the body has already led to some inexpensive and widely used thermometry applications. Replacing rectal thermometers with infrared ear thermometers in pediatric clinics is arguably one of the most widely used and appreciated applications of electromagnetics today. Numerous noncontact temperature measurement applications in medicine use infrared thermometry. Since temperature rise is often associated with injury, the use of infrared thermometry for noninvasive diagnosis of soft tissue injury has already gained widespread use in veterinary medicine. (Probably the main reason that it has found application in animals is its low cost.) Additional applications of infrared thermometry will undoubtably continue to expand.

Applications of visible light in medicine originated with optical imaging and with the detection of color and hue in conjunction with conditions that change the color. Determination of blood oxygen level is one such application. Other applications are likely, following the desire to automate tests that were previously done manually. Image recognition software originally developed for photographic and radar applications can be used for reading microscope slides, x-ray, CT, and MRI images, for instance. This application is not yet widespread, but is likely to be enhanced in the future.

Applications of lasers and fiber optics in medicine have expanded at a fast rate. They are used for cutting, ablation, and diagnostics. They have been enabling technology in minimally invasive surgery (laparoscopy). The tools and methods for this type of surgery continue to shrink in size yet gain in effectiveness. Most gall bladder and orthopedic surgeries are now performed laparoscopically. Eye surgery and surgical vision correction are often done with laser tools. Even some heart surgeries, such as repair of septal defects, are done with minimally invasive surgical tools. Diagnostic tools for urinary, esophageal, and intestinal tract disorders utilize fibers and lenses to guide visible light to see deep into the body. These tools are also likely to expand in capability and reduce in size, enabling diagnosis on a more localized scale.

Finally, x-ray applications are ubiquitous in medicine. These applications were not described in detail in this book because of the significant difference in the behavior of x-rays and other electromagnetic waves. X-rays are of such high frequency that they have enough energy to break cell bonds. Other electromagnetic waves do not. In this regard, x-rays fall into a very different category from the other waves described in this book.

6.7 Conclusion

All of the applications of electromagnetics are based on the ability to get the power into and out of the body and the ability to focus it on the region of interest. These capabilities are controlled by the laws of physics that control attenuation, diffraction, and scattering of electromagnetic waves. But they are also very much impacted by our ability to be creative. Use of invasive systems, for instance, brings the transmitters and receivers much closer to the region of interest and provides more localized fields. The use of signal and image processing allows us to combine many less localized measurements to determine very detailed, highly localized images of the body. This kind of development is spurring the next generation of electromagnetic devices. Sharper images, new and improved diagnostics, more effective treatments, and enhanced understanding of how a living organism functions are potential benefits of bioelectromagnetics.

Appendix A: Electrical Properties of the Human Body

The electrical properties of human tissues (relative permittivity ε_r' and conductivity σ_{eff}) control the propagation, reflection, attenuation, and other behavior of electromagnetic fields in the body that are described within this book. These properties depend strongly on the tissue type and the frequency of interest. Temperature, blood or fluid perfusion, and individual differences are second-order effects that are normally not considered. The body is so weakly magnetic that generally μ_r is assumed to be 1, except for magnetic resonance imaging and spectroscopy applications where a very large magnetic field is used.

Figure A.1 shows the electrical properties of two very extreme tissues in the body—muscle and fat—as a function of frequency. Muscle has very high salinity and high water content, and fat is low water content. Muscle is a good conductor (higher conductivity σ_{eff}), and fat is a poor conductor (good insulator, lower conductivity). The loss tangent $\tan\delta = \varepsilon''/\varepsilon'$ tells which component of the electrical properties dominates the effect on the field. At low frequencies, the conductivity of the tissue dominates the behavior of the field ($\varepsilon_r'' = \sigma_{eff}/\varepsilon_0\omega$, and ω is small), and at high frequencies, the relative permittivity ε_r' tends to dominate. The values of conductivity (σ) in this appendix are the effective conductivity (σ_{eff}) as defined in Section 1.14, which is consistent with most other tables of conductivity you will find in the literature. Table A.1 shows the electrical properties of several different tissues in the body at 433 MHz, which is a commonly used frequency for Industrial Scientific Medical (ISM) applications. A common approximation is that the body can be modeled using average properties of 2/3 muscle, which means that both ε_r and σ at the frequency of interest are multiplied by 2/3. This is suitable for addressing global questions such as total power absorbed in the body, but is generally not suitable for evaluating near-field effects such as peak specific absorption rate (SAR).

The electrical properties of the body (ε_r and σ_{eff}) control the wavelength and attenuation according to the equations given in Section 1.14. The attenuation of the field is calculated as $e^{-\alpha z}$, where z is the distance the wave must propagate through that tissue. At 433 MHz, 69% of the field is transferred through 10 cm of fat, and 11% is transferred through 10 cm of muscle. The higher-water-content (higher-conductivity) tissues have more attenuation. The wavelength is calculated from $2\pi/\beta$ (meters). The wavelength at 433 MHz in fat is 30 cm, and in muscle is 8 cm. A typical rule of thumb is that an antenna should be half a wavelength, which would be 4 cm in muscle. While this still seems too large for most implantable devices, specialized antenna designs can achieve performance in the body at this frequency.

Electromagnetic measurements such as assessment of cellular telephones, evaluation of the performance of telemetry (communication) devices implanted in the body, or other measurement applications often require body-simulant materials. These can be solid, semisolid, or (most commonly) liquid materials that have electrical properties that mimic those of human tissues.

High-water-content tissue stimulant materials (for brain, muscle, etc.) are commonly made from water, polyethylene powder, and sodium chloride (NaCl). TX-150 may be used as a gelling agent in order to create a semisolid material. This type of material has been used from 13.56 to 2450 MHz (Guy, 1971; Chou et al., 1984). An example of this type of

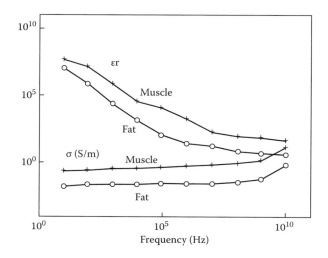

FIGURE A.1

Electrical properties of muscle and fat. (From Gabriel, C. 1996. *Compilation of the dielectric properties of body tissues at RF and microwave frequencies,* Final technical report, Occupational and Environmental Health Directorate Radiofrequency Radiation Division, Brooks Air Force Base, TX.)

TABLE A.1

Electrical Properties of Tissues at 433 MHz

Tissue	ε_r	σ_{eff} (S/m)	Tissue	ε_r	σ_{eff} (S/m)
Air (vacuum)	1	0	Lens cortex	52.75	0.6742
Aorta	49.15	0.7395	Lens nucleus	38.76	0.38
Bladder	17.67	0.3128	Liver	50.34	0.68
Blood	57.3	1.72	Lung deflated	52.83	0.7147
Bone (cancellous)	21.08	0.02275	Lung inflated	21.58	0.3561
Bone (cortical)	13.77	0.1032	Muscle	64.21	0.9695
Bone (marrow)	5.137	0.03575	2/3 muscle	42.81	0.6463
Breast fat	5.62	0.04953	Nerve	35.7	0.500
Cartilage	43.64	0.65	Ovary	51.55	1.033
Cerebellum	52.9	0.91	Skin (dry)	42.48	0.5495
Cerebrospinal fluid	68.97	2.32	Skin (wet)	51.31	0.72
Cervix	44.17	1.020	Small intestine	74.1	2.053
Colon	60.88	0.96	Spleen	60.62	1.041
Cornea	54.4	1.070	Stomach	74.55	1.120
Dura	51.03	0.8	Tendon	50.53	0.7554
Eye tissues	57.69	1.010	Testes	65.2	1.137
Fat	5.028	0.04502	Thyroid	60.02	0.8183
Gall bladder	60.06	1.035	Tongue	58.79	0.8993
Gall bladder bile	76.55	1.613	Trachea	42.93	0.673
Gray matter	54.27	0.8775	Uterus	64.73	1.117
Heart	60.74	0.9866	Vitreous humour	66.16	0.3931
Kidney	57.3	1.152	White matter	39.84	0.5339

Source: Gabriel, C., *Compilation of the dielectric properties of body tissues at RF and microwave frequencies,* Final technical report, Occupational and Environmental Health Directorate Radiofrequency Radiation Division, Brooks Air Force Base, TX, 1996.

phantom is shown in Figure 5.1. Polyacrylamide gel has also been used; however, this material is more difficult to fabricate and degrades rapidly if exposed to air (which is typically the case during testing) (Bini et al., 1984). Simple water and gelatin with sugar and salt to control the electrical properties have also been used. Although inexpensive and easy to create, this material is more like a thick liquid than a gel, and therefore is limited to homogenous phantoms (Marchal et al., 1989; Sunaga et al., 2003). A solid phantom created by Chang et al. (2000) uses a solid conductive plastic made of polyethyl methacrylate and carbon black. One of the advantages of this material is that it can be cast into any shape.

A low water content phantom (for fat and bone) has been created from Laminac 4110 (a polyester resin), acetylene black, and aluminum powder. This solid material has been used from 100 MHz to 10 GHz (Guy, 1971; Cheung and Koopman, 1976.) A semisolid dough (made of flour, oil, and saline) has been tested by Lagendijk and Nilsson (1985), and silicone rubber with carbon fiber was tested by Nikawa et al. (1996).

References

Bini, M. G., Ignesti, A., Millanta, L., et al. 1984. The polyacrylamide as a phantom material for electromagnetic hyperthermia studies. *IEEE Trans. Biomed. Eng.* 31:275–77.

Chang, J. T., Fanning, M. W., Meaney, P. M., et al. 2000. A conductive plastic for simulating biological tissue at microwave frequencies. *IEEE Trans. Electromagn. Compat.* 42:76–81.

Cheung, A. Y., and Koopman, D. W. 1976. Experimental development of simulated biomaterials for dosimetry studies of hazardous microwave radiation. *IEEE Trans. Microw. Theory Tech.* 24:669–73.

Chou, C.-K., Chen, G.-W., Guy, A. W., et al. 1984. Formulas for preparing phantom muscle tissue at various radiofrequencies. *Bioelectromagnetics* 5:435–41.

Gabriel, C. 1996. *Compilation of the dielectric properties of body tissues at RF and microwave frequencies*, Final technical report. Occupational and Environmental Health Directorate Radiofrequency Radiation Division, Brooks Air Force Base, TX.

Guy, A. W. 1971. Analysis of electromagnetic fields induced in biological tissue by thermographic studies on equivalent phantom models. *IEEE Trans. Microw. Theory Tech.* 19:189–217.

Lagendijk, J. J. W., and Nilsson, P. 1985. Hyperthermia dough: A fat and bone equivalent phantom to test microwave/radiofrequency hyperthermia heating systems. *Phys. Med. Biol.* 30:709–12.

Marchal, C., Nadi, M., Tosser, A. J., et al. 1989. Dielectric properties of gelatin phantoms used for simulations of biological tissues between 10 and 50 MHz. *Int. J. Hyperthermia* 5:725–32.

Nikawa, Y., Chino, M., and Kikuchi, K. 1996. Soft and dry phantom modeling material using silicone rubber with carbon fibre. *IEEE Trans. Microw. Theory Tech.* 44:1949–53.

Sunaga, T., Ikehira, H., Furukawa, S., et al. 2003. Development of a dielectric equivalent gel for better impedance matching for human skin. *Bioelectromagnetics* 24:214–17.

Surowiec, A., Shrivastava, P. N., Astrahan, M., et al. 1992. Utilization of a multilayer polyacrylamide phantom for evaluation of hyperthermia applicators. *Int. J. Hyperthermia* 8:795–807.

Appendix B: Definition of Variables

TABLE B.1

Definitions of Variables

Variable	Name	Units	Notes	Type	How to Say It	Section
α	Attenuation constant	Nepers/ meter	Np/m	Scalar	Alpha	1.14
B	Magnetic flux density	Tesla	T	Vector		1.3
B	Magnitude of the magnetic flux density	Tesla	T	Scalar		1.3
β	Propagation constant	Radians/ meter	rad/m	Scalar	Beta	1.11, 1.14
c or c_o	Speed (velocity of propagation) of light	Meters/ second	$= 3 \times 10^8$ m/s	Constant		1.11
c	Specific heat	Joules/ degree-kg	J/°C-kg	Relative constant		1.16
d	Distance	Meters	m	Scalar		
d	Size of diffraction slit	Meters	m	Scalar		3.8.1
d_i	Distance to the image	Meters	m	Scalar		4.2.4
d_o	Distance to the object	Meters	m	Scalar		4.2.4
D	Electric flux density	Coulombs/ square meter	C/m^2	Vector		1.7
D	Magnitude of the electric flux density	Coulombs/ square meter	C/m^2	Scalar		1.7
D	Diameter of lens	Meters	m	Scalar		4.2.3
D	Maximum dimension of antenna	Meters	m	Scalar		3.7
$\partial/\partial t$	Change with respect to time	Per second	1/s			1.4
E	Electric field	Volts/meter	V/m	Vector		1.2
E	Magnitude of the electric field	Volts/meter	V/m	Scalar		1.2
$\tilde{\mathbf{E}}$	Phasor (frequency domain) electric field	Volts/meter	V/m	Complex valued vector		1.14
$\mathbf{E}_{inc}$, $\mathbf{E}_i$	Incident electric field	Volts/meter	V/m	Vector		5.5.2, 3.3
$\mathbf{E}_{Eint}$	Internal **E** field produced by $\mathbf{E}_{inc}$	Volts/meter	V/m	Complex valued vector		5.5.2
$\mathbf{E}_{Hint}$	Internal **E** field produced by $\mathbf{H}_{inc}$	Volts/meter	V/m	Complex valued vector		5.5.2
E_{rms}	RMS electric field	Volts/meter	V/m	Scalar		1.40
$\mathbf{E}_{int}$	Internal electric field	Volts/meter	V/m	Vector		2.2

(continued on next page)

257

TABLE B.1 (continued)

Definitions of Variables

Variable	Name	Units	Notes	Type	How to Say It	Section
e	Constant	Unitless	2.718	Scalar		1.13, 1.14, 4.4.2
eV	Voltage relative to potential of an electron	Electron volts	eV	Scalar		4.7
ε	Permittivity	Farads/ meter	$= \varepsilon_0\, \varepsilon_r$	Scalar	Epsilon	1.6, 1.4
ε_0	Permittivity of free space	Farads/ meter	8.854×10^{-12} F/m	Constant	Epsilon naught	1.6
ε_r	Relative permittivity	Unitless	Range typically 1–80	Scalar	Epsilon r	1.6
ε_{bound}	Complex permittivity from bound charges	Farads/ meter	$= \varepsilon_{b,real} - j\varepsilon_{b,imag}$	Scalar		1.14
$\varepsilon_{b,real}$	Real part of complex permittivity from bound charges	Farads/ meter		Scalar		1.14
$\varepsilon_{b,imag}$	Imaginary part of complex permittivity from bound charges	Farads/ meter		Scalar		1.14
ε''	Imaginary part of the complex permittivity	Unitless	$= \varepsilon_{eff}/\omega\varepsilon'$	Scalar		1.14
ε'	Real part of the complex permittivity	Unitless	$= \varepsilon_0\varepsilon_r$	Scalar		1.14
F	Force	Newtons	N	Vector		1.2
f	Frequency	Hertz = 1/second	Hz = 1/s	Scalar		1.9
f	Focal length	Meters	m	Scalar		4.2.3
f# or f/ or f	f-number of lens	Unitless		Scalar		4.2.3
ϕ	Phase of a sine wave	Degrees or radians		Scalar	Phi	1.9
H	Magnetic field	Amperes/ meter	A/m	Vector		1.7
H	Magnitude of the magnetic field	Amperes/ meter	A/m	Scalar		1.7
$\mathbf{H}_{inc}$, $\mathbf{H}_I$	Incident magnetic field	Amperes/ meter	A/m	Vector		5.5.2, 3.3
$\tilde{\mathbf{H}}$	Phasor (frequency domain) magnetic field	Amperes/ meter	A/m	Complex valued vector		1.14
Hrms	RMS magnetic field	Amps/ meter	A/m	Scalar		1.40
I	Current	Amperes (amp)	A	Scalar		1.2
$\tilde{I}$	Phasor current	Amperes (amp)	A	Complex scalar		1.13
J	Current density	Amps/ square meter	A/m^2	Vector		1.5

TABLE B.1 (continued)

Definitions of Variables

Variable	Name	Units	Notes	Type	How to Say It	Section
$J_{c,eff}$, J_c	Conduction current density	Amps/ square meter	A/m^2	Vector		1.14
J_d	Displacement current density	Amps/ square metr	A/m^2	Vector		1.15
j or i	$\sqrt{-1}$	Unitless				1.13
k	Direction of propagation	Unitless		Unit vector		3.2
λ	Wavelength	Meters	m	Scalar	Lambda	1.6, 1.14
λ_d	Wavelength in dielectric	Meters	m	Scalar	Lambda	3.3.2
λ_{co}	Cutoff wavelength	Meters	m	Scalar	Lambda	3.5.3.2
M	Magnification factor	Unitless	$= d_i/d_o$	Scalar		4.2.4
n	Index of refraction	Unitless	$= \sqrt{\varepsilon_r}$	Scalar		4.1
NA	Numerical aperture	Unitless		Scalar		4.3.1
P	Power	Watts	W	Scalar		1.16
p	Pressure of ultrasound wave			Scalar		3.9
Q	Charge	Coulombs	C	Scalar		1.2
Q	Quality factor of a cavity	Unitless		Scalar		3.6
R	Resistance	Ohms	Ω	Scalar		1.2
ρ	Mass density	Kilogram/ cubic meter	kg/m^3	Scalar	Rho	1.16
ρ	Charge density	Coulombs/ cubic meter	C/m^3	Scalar	Rho	1.4
ρ	Reflection coefficient	Unitless		Scalar		3.5, 4.2.2
rms	Root mean square					1.10
R_0	Radius of curvature (of lens)	Meters	m	Scalar		4.2.3
SAR	Specific absorption rate	Watts/kg	W/kg	Scalar	Sar	1.16
σ	Conductivity	Siemens/ meter	$S/m = 1/\Omega\text{-m}$	Scalar	Sigma	1.6, 1.14, 1.11
σ_c	Conductivity from free electrons	Siemens/ meter	$S/m = 1/\Omega\text{-m}$	Scalar		1.14
σ_{eff}	Effective conductivity that represents loss from free electron motion and displacement of bound charges	Siemens/ meter	$S/m = 1/\Omega\text{-m}$	Scalar		1.14
δ	Skin depth	Meters	$\delta = 1/\alpha$	Scalar		1.14, 3.4.1

(continued on next page)

TABLE B.1 (continued)

Definitions of Variables

Variable	Name	Units	Notes	Type	How to Say It	Section
T	Period of a sine wave	Seconds	s	Scalar		1.9
TE	Transverse electric			Definition		3.5
TM	Transverse magnetic			Definition		3.5
TEM	Transverse electric and magnetic			Definition		3.5
ΔT	Temperature rise	Degrees		Scalar		1.16
t	Time	Seconds	s	Scalar		1.14, 3.3
$\tan \delta$	Loss tangent (dissipation factor)	Unitless	$= \varepsilon''/\varepsilon'$	Scalar		1.14
θ_B	Brewster's angle	Radians or degrees		Scalar		4.2.2
θ_{ic}	Critical angle (total internal reflection)	Radians or degrees		Scalar		3.3.2, 4.3
θ_i	Angle of incidence	Radians or degrees		Scalar		3.3.2
θ_t	Angle of transmission	Radians or degrees		Scalar		3.3.2
θ_d	Half angle of divergence	Radians or degrees		Scalar		3.8.1
θ_n	Half angle nth diffraction peak	Radians or degrees		Scalar		3.8.2
ϕ_r	Half angle of radiated cone of light	Radians or degrees		Scalar		4.3.1
u	Velocity of ultrasound wave	Meters/ second	m/s	Scalar		3.9
μ	(Magnetic) permeability	Henrys/ meter	$= \mu_0 \mu_r$	Scalar	mu	1.5, 1.14
μ_0	Permeability of free space	Henrys/ meter	$4\pi\,10^{-7}$ H/m	Constant	mu naught	1.6
μ_r	Relative permeability	Unitless		Scalar	mu r	1.6
$\mu_{complex}$	Complex permeability	Henrys/ meter	$= \mu' - j\,\mu''$	Complex scalar		1.14
μ'	Real part of complex permeability	Henrys/ meter	$= \mu_0\mu_r{}'$	Scalar		1.14
μ''	Imaginary part of complex permeability	Henrys/ meter	$= \mu_0\mu_r{}''$	Scalar		1.14
$\mu_r{}' - j\,\mu_r{}''$	Complex relative permeability	Unitless		Scalar	mu r	1.14
$\mu_s{}'$	Scattering coefficient	1/meter	1/m	Scalar		4.6.1
V	Voltage	Volts	V	Scalar		1.2
$\tilde{V}$	Phasor voltage	Volts	V	Complex scalar		1.13
V_{inc}	Incident voltage	Volts	V	Complex scalar		3.5
V_{ref}	Reflected voltage	Volts	V	Complex scalar		3.5
$|V|_{max}$	Maximum value of the voltage standing wave	Volts	V	Scalar		3.5

TABLE B.1 (continued)

Definitions of Variables

Variable	Name	Units	Notes	Type	How to Say It	Section
$|V|_{min}$	Minimum value of the voltage standing wave	Volts	V	Scalar		3.5
v_p	Phase velocity or velocity of propagation	Meters/second	m/s	Scalar		1.11
VSWR	Voltage standing wave ratio	Unitless		Scalar		3.5
Δv	Small unit of volume	Cubic meters	m^3			1.16
ω	Angular frequency	Radians/second	$rad/s = 2\pi f$	Scalar	Omega	1.9
$w(z)$	Waist (of laser beam)	Meter		Scalar		4.4.2
w_0	Waist (of laser beam)	Meter	$= w(z = 0)$	Scalar		4.4.3
W_{em}	EM power deposition	Watts	W	Scalar		1.16
W_m	Metabolic heating rate	Watts	W	Scalar		1.16
W_c	Thermal dissipation by conduction	Watts	W	Scalar		1.16
W_b	Thermal dissipation by blood flow	Watts	W	Scalar		1.16
z	Distance	Meter	m	Scalar		1.14, 3.3
Z	Complex impedance	Ohms	Ω	Complex scalar		1.13, 3.2, 3.9
Z_L	Complex impedance of load	Ohms	Ω	Complex scalar		3.5
Z_0 (some books call this η or η_0)	Characteristic impedance	Ohms	Ω	Complex scalar (often real scalar)		3.5
z_R	Rayleigh range	Meter		Scalar		4.4.3

TABLE B.2

Multiplication Factors for Units

Unit	Unit Label	Multiplication Factor
Terra	T	$\times 10^{12}$
Giga	G	$\times 10^9$
Mega	M	$\times 10^6$
Kilo	k	$\times 10^3$
Centi	c	$\times 10^{-2}$
Milli	m	$\times 10^{-3}$
Micro	μ	$\times 10^{-6}$
Nano	n	$\times 10^{-9}$

Example: 5 MHz (megahertz) = 5×10^6 Hz.

Appendix C: Decibels

Electromagnetic fields often vary wildly in strength. For instance, magnetic resonance imaging (MRI) machines may create very large external magnetic fields to cause the body's hydrogen atoms to align with the field. When the field is turned off and the atoms return to their random state, the fields they generate are very small. In order to simplify the description and calculation of these very different field strengths, decibel units are used. Another significant advantage of using decibels is that they simplify the mathematics of the linear systems that are also often found in electromagnetics.

A decibel (dB) value is calculated this way:

$$\text{Anything in dB} = 10 * \log_{10} (\text{anything in linear})$$

$$\text{Power in dB} = 10 * \log_{10} (\text{power in watts})$$

A simple relationship between power and voltage (or field) in decibels also exists. Recall that power = V^2/R or $|E|^2/Z$. Then:

$$\text{Power in dB} = 10 * \log_{10} (\text{power in watts})$$

$$= 10 * \log_{10} (V^2/R)$$

$$= 20 * \log_{10} (\text{voltage}/\sqrt{R})$$

Specific names have been given to decibel units for different orders of magnitude of the power:

$$\text{Power (dBW, also often just called dB)} = 10 * \log_{10} (\text{power in watts})$$

$$\text{Power (dBm)} = 10 * \log_{10} (\text{power in milliwatts})$$

$$\text{Power (dBµ)} = 10 * \log_{10} (\text{power in microwatts})$$

Decibels can simplify electromagnetic math to allow you to do it in your head. For example, if you double the strength of something, you just add 3 dB. Table C.1 gives several common functions and their dB equivalent.

TABLE C.1

Common dB Factors

Math Function	dB Function
1	0 dB
*2	+3 dB
/2	−3 dB
*10	+10 dB
*100	+20 dB
/10	−10 dB
/100	−20 dB

TABLE C.1 (continued)

Common dB Factors

Math Function	dB Function
Example:	
1 watt	0 dB
2 watts	0 + 3 dB = 3 dB
20 watts	= 3 dB + 10 = 13 dB

Decibels are also used to simplify the calculation of the effect of various gains and losses in a system.

Suppose we are given the power transmitted P_{TX} and want to find the power received P_{RX}. This is a multiplicative function of gains and losses:

$$P_{RX} = P_{TX} * gain/loss$$

If there are numerous gains and losses in a system, all of the gains are multiplied, and all of the losses are divided. The mathematics typically gets too difficult to do in your head. But logarithmic functions (such as decibels) convert multiplication and division to addition and subtraction. Now the math is often simple enough to do in your head!

$$P_{RX} (dB) = P_{TX} (dB) + gain (dB) - loss (dB)$$

An important detail in this type of calculation is to realize that losses and gains are relative values. They are ratios and do not have units. Thus, they are always simply dB. The power terms in these equations do have units. They may be in watts, milliwatts, and so forth. Thus, they must use the naming convention for power described above. Then there will be one decibel with a unit (such as dBm) on the right-hand side of the equation. The dB with a unit on the left-hand side will be the same as that on the right-hand side (in this case dBm):

$$P_{RX} (dBm) = P_{TX} (dBm) + gain (dB) - loss (dB)$$

Index